HANDBOOK OF ISOELECTRIC FOCUSING AND PROTEOMICS

This is Volume 7 of

SEPARATION SCIENCE AND TECHNOLOGY

A reference series edited by Satinder Ahuja

HANDBOOK OF ISOELECTRIC FOCUSING AND PROTEOMICS

Edited by

David Garfin

Biotechnology Consultant
Kensington, California

Satinder Ahuja

Ahuja Consulting
Calabash, North Carolina

ELSEVIER
ACADEMIC
PRESS

Amsterdam – Boston – Heidelberg – London – New York – Oxford – Paris – San Diego – San Francisco – Singapore – Sydney – Tokyo

ELSEVIER B.V.
Radarweg 29
P.O. Box 211, 1000 AE
Amsterdam,
The Netherlands

ELSEVIER Inc.
525 B Street, Suite 1900 San Diego, CA 92101-4495
USA

ELSEVIER Ltd
The Boulevard,
Langford Lane
Kidlington,
Oxford OX5 1GB
UK

ELSEVIER Ltd
84 Theobalds Road
London WC1X 8RR
UK

First edition 2005

ISBN: 0-12-088752-5

♾ The paper used in this publication meets the requirements of ANSI/NISO Z39.48-1992 (Permanence of Paper).

Printed in The Netherlands.

CONTENTS

9. Capillary Isoelectric Focusing

TIM WEHR

10. Free-Flow Isoelectric Focusing

PETER J.A. WEBER, GERHARD WEBER, CHRISTOPH ECKERSKORN, ULRICH SCHNEIDER, AND ANTON POSCH

11. Isoelectric Focusing and Proteomics

MELANIE Y. WHITE AND STUART J. CORDWELL

12. Chromatofocusing

DAVID ANDERSON

13. Alternative Electrofocusing Methods

CORNELIUS F. IVORY

PREFACE

Separation science and technology play a major role in protein biochemistry. Of all the methods available for fractionating proteins, none offers higher resolving power than isoelectric focusing (IEF). It can be used in an analytical or preparative mode to separate proteins on the basis of their unique isoelectric points. This technique also lends itself to multidimensional separations because it is compatible with a large variety of other separation methods, especially chromatography and gel electrophoresis. IEF has much to offer protein chemists, and it is generally included in the repertoire of standard laboratory methods of anyone studying proteins at any level.

The emergence of proteomics as a mature discipline has brought about a renewed interest in IEF, as it is a rapid and reliable means for protein fractionation. The basic concept of proteomics, massively parallel protein analysis, grew from considerations of two-dimensional polyacrylamide gel electrophoresis (2-DE), where hundreds to thousands of proteins are separated and visualized in a single entity. Two-dimensional gels are an integral part of proteomics research, where they serve as the main separation tool. Since the first-dimension separation of 2-DE is by IEF, the challenge of performing successful 2-DE essentially requires the achievement of good IEF fractionations. This fact has provided the incentive to develop a text that considers IEF in a proteomics context.

This *Handbook of Isoelectric Focusing and Proteomics* is intended to be a single source for relevant information on IEF. The book addresses many of the salient features of IEF and has been written by a panel of experts in the field. The book expands on a chapter dealing with IEF in the *Handbook of Bioseparations* (S. Ahuja, Editor, Academic Press, San

Diego, 2000), broadening the scope of that material to cover more thoroughly topics that should be of interest to a wide range of readers.

We have arranged the subjects in a logical order, beginning with the history of IEF through to alternative forms of focusing. The chapters are independent of one another, and each chapter includes the authors' special insights and practical tips. Chapter 1 provides a broad overview of IEF and proteomics. The history of IEF goes back to the early days in the electrophoretic analysis of proteins. Chapter 2 discusses the evolution of the technique and includes photographs of some of the pioneers in the field. Theoretical descriptions of IEF covered in Chapter 3 range from simple differential equations to sophisticated computer models. Generation of pH gradients is the key element in IEF; Chapter 4 is devoted to the synthetic molecules used to generate them. The practical aspects of slab gel IEF are presented in Chapter 5. This chapter is designed to be useful to novices and to experienced users as well, and it provides many useful tips and insights. Chapter 6 describes the 2-DE methodology, with an emphasis on the second-dimension gel. The key to successful IEF and, by extension, successful 2-DE, lies in sample preparation. Since protein mixtures vary widely in their properties, no single sample preparation procedure is universally applicable to all samples. Chapter 7 covers many important considerations in sample preparation. The detection methods used for visualizing proteins in IEF gels are explained in Chapter 8. The important topics of image acquisition and analysis are also covered in this chapter.

All of the relevant techniques of capillary IEF, along with discussions of their advantages and disadvantages, are presented in Chapter 9. Methods for preparative IEF are described in several chapters. Because of the high sensitivity offered by mass spectrometers, applications of capillary IEF, which generates low quantities of proteins, can be considered preparative. Chapter 10 covers free-flow electrophoresis (FFE), a preparative technique with a great deal of promise.

The role of IEF in 2-D gel-based proteomics is addressed in Chapter 11. The pros and cons of 2-DE proteomics are discussed, including the well-known under-representation of hydrophobic or membrane-associated proteins, highly alkaline proteins, high- and low-molecular-weight proteins, and lower abundance proteins in relation to high abundance "housekeeping" proteins. Several methods based on IEF that are designed to overcome these problems are discussed, including preliminary fractionation by IEF and modifications to the standard first-dimension IEF procedures.

The final two chapters diverge from descriptions of purely IEF methods in order to highlight related, alternative approaches. Chromatofocusing, discussed in Chapter 12, is a chromatographic technique, rather than an electrophoretic one, and is similar to IEF at very basic levels. This technique is more of a preparative method than an analytical one, since the pI values obtained are only approximate. It is most often used as part of a protein purification scheme. Chapter 13 considers IEF in the context of related equilibrium gradient methods. In contrast to IEF, the set of alternative methods that are described are in the conceptual or very-early-development phase. They point to some of the possible routes that separation technology can take in the search for efficient techniques for fractionating and purifying proteins.

We thank the authors for their excellent contributions. This book should enable anyone working with proteins, from the level of graduate student to senior researcher, to find something useful in the text, as it describes underlying principles of IEF and its many applications, including proteomics.

We are sure the readers will be fascinated by Pier Giorgio Righetti's personal reminiscences on the early development of this technique in the next few pages.

David Garfin
Satinder Ahuja

CONTRIBUTORS

Numbers in parentheses indicate the page on which the authors' contributions begin.

Satinder Ahuja (1) Ahuja Consulting, 1061 Rutledge Court, NW, Calabash, NC 28467 USA

David Anderson (265) Department of Chemistry, Cleveland State University, 2121 Euclid Avenue, Cleveland, OH 44115 USA

J. C. Baygents (41) Department of Chemical & Environmental Engineering, The University of Arizona, Tucson, AZ 85721 USA

Tom Berkelman (69) Life Science Group, Bio-Rad Laboratories, 6000 James Watson Drive, Hercules, CA 94547 USA

Stuart J. Cordwell (247) Australian Proteome Analysis Facility, Level 4, Building F7B, Macquarie University, North Ryde, Sydney, NSW 2109, Australia

Christoph Eckerskorn (211) FFEWeber GmbH, IZB, Building 6, Am Klopferspitz 19, D-82152 Planegg/Munich, Germany

David Garfin (1) Biotechnology Consultant, 112 Kenyon Avenue, Kensington, CA 94708 USA

Ben Herbert (147) Proteome Systems Ltd., 1/35-41 Waterloo Road, North Ryde, Sydney, NSW 2113, Australia

Cornelius F. Ivory (297) Department of Chemical Engineering, Washington State University, Pullman, WA 99164-2710 USA

Ferenc Kilár (13) Department of Analytical Chemistry, Faculty of Sciences, Ifjúság útja 6, 7624 Pécs and Institute of Bioanalysis, Faculty of Medicine, University of Pécs, Szigeti út 12, H-7643 Pécs, Hungary

Michael T. McDowell (123) Pfizer Global Research and Development, Molecular Technologies, Ann Arbor, MI 48105 USA

Mark P. Molloy (123) Pfizer Global Research and Development, Molecular Technologies, Ann Arbor, MI 48105 USA

Wayne F. Patton (165) PerkinElmer LAS, Building 100-1, 549 Albany Street, Boston, MA 02118 USA

Anton Posch (211) Proteoconsult, Assinger Strasse 2a, D-85567 Grafing/Munich, Germany

Pier Giorgio Righetti (xvii) Department of Agricultural and Industrial Biotechnologies, University of Verona, Strada Le Grazie No. 15, I-37134 Verona, Italy

P. A. Safier (41) Department of Chemical & Environmental Engineering, The University of Arizona, Tucson, AZ 85721 USA

Ulrich Schneider (211) FFEWeber GmbH, IZB, Building 6, Am Klopferspitz 19, D-82152 Planegg/Munich, Germany

T. L. Sounart (41) Sandia National Laboratories, Albuquerque, NM 87185-1411 USA

Ákos Végvári (13) Department of Analytical Chemistry, Faculty of Sciences, Ifjúság útja 6, 7624 Pécs and Institute of Bioanalysis, Faculty of Medicine, University of Pécs, Szigeti út 12, H-7643 Pécs, Hungary

Gerhard Weber (211) FFEWeber GmbH, IZB, Building 6, Am Klopferspitz 19, D-82152 Planegg/Munich, Germany

Peter J. A. Weber (211) FFEWeber GmbH, IZB, Building 6, Am Klopferspitz 19, D-82152 Planegg/Munich, Germany

Tim Wehr (181) Life Science Group, Bio-Rad Laboratories, 6000 James Watson Drive, Hercules, CA 94574 USA

Reiner Westermeier (93) Amersham Biosciences Europe GmbH, Munzinger Strasse 9, D-79111 Freiburg, Germany

Melanie Y. White (247) Australian Proteome Analysis Facility, Level 4, Building F7B, Macquarie University, North Ryde, Sydney, NSW 2109, Australia

INCIDENTS OF TRAVEL IN IEF AND IPGS*

PIER GIORGIO RIGHETTI

University of Verona, Italy

That winter of 1942 must have been the most dreadful one in the life of poor Harry Svensson. As a young pupil, under the iron fist of Dom Arne Tiselius, he was forced to spend endless days and nights trying to make a dream of his boss come true, namely, the creation of a pH gradient by the stationary electrolysis of a salt solution. Among the innumerable problems encountered, one became immediately apparent: the ionic constituents were completely swept to the opposite-charge electrode. Although equilibrium between ion transport and backdiffusion ensued, at neutral pH, midway in the apparatus where a region of water almost completely devoid of ions formed, ohmic resistance increased enormously with the result that the liquid almost boiled at this point.

Luckless Harry was chained to the bench adding salt dropwise from a burette, trying to fight the conductivity minimum. Add to this the fact that it was the peak of World War II: Norway and Denmark were occupied by Germany; Finland was oppressed by neighboring Russia; and Sweden ducked low under a declaration of neutrality, yet under quite miserable conditions. Harry dreamed of escaping this agonizing life and paltry winter, repairing to England, being drafted in the army and sent to the African desert to fight the Italian–German coalition. At least he would have enjoyed sunny days and warm weather!

As luck would have it, Tiselius relaxed his iron grip and let Harry present his Ph.D. thesis in 1946 (it was on "Electrophoresis by the Moving Boundary Method").[1] This was no ordinary thesis, mind you, for Harry's mentors were The Svedberg, a 1926 Nobel laureate, and Tiselius, a future Nobelist (1948).

Hapless Harry must have been obsessed for the remainder of his life by this idea of creating pH gradients. Already in 1956 he dreamed of a hypothetical biprotic ampholyte having two intrinsic p*K* values both equal to 7.0. If it were ever to exist, a solution of it in pure water would be a neutral buffer containing one-fourth of the amount in the cationic and one-fourth in the anionic form. A superb "carrier ampholyte" (CA), indeed, which would offer a substantial conductivity where it was most needed, i.e., in that terrible conductivity "Grand Canyon" located at

*A paraphrase of the surely more famous book "*Incidents of Travel in Central America, Chiapas and Yucatan*, by John L. Stephens, 2 volumes, New York, 1841.

pH 7, but which would still be isoelectric and immobile in the electric field.[2] Although such a "super-buffer" could not possibly exist, 3 years later, during a leave of absence at Caltech, chez Linus Pauling, Svensson found a real chemical that came rather close to that: histidylhistidine, with p*K* values at 6.8 and 7.8, thus isoelectric at pH 7.3. It was used with hemoglobins in the first experiments.[3]

Once back home, as a freelance at the Karolinska Institute, Harry worked hard on the theory of isoelectric focusing (IEF) and laid down its theoretical foundations in a couple of, by now, classic papers.[4,5] His secret battle cry as he waged his underground guerrilla warfare against the Maestro's mammoth instrumentation was: "no more moving boundaries!" He had a point. It had been foolish to try to create stable and stationary pH gradients in the presence of an electric field with non-amphoteric compounds. These would only vacate the grounds and leave an empty trail in their wake with no soldiers to guard the battlefield. All buffers had to be amphoteric and, in addition, they had to have decent buffering power as ensured by not too large Δp*K* values. Only in this way would the zones of isoelectric "carrier ampholytes" form a continuous chain, as the electric field would tie them to their isoelectric zones while diffusion would cause them to broaden just enough to penetrate the neighboring ampholyte zones, thus simultaneously ensuring buffer capacity and conductivity. The result was not just a few moving boundaries, *à la Tiselius*, but a horde of stationary boundaries, each one standing guard against local pH changes. It was too bad that this army was made up of barely a handful of soldiers, hardly able to cover the grounds in the pH 3–10 interval. Nonetheless, Svensson published, in 1964, some remarkable color pictures of unique hemoglobin (Hb) separations in his 110 mL focusing column stabilized by a sucrose density gradient. In these separations, Harry used protein lysates, notably of casein, albumin, Hb, and whole blood, as background carrier ampholytes. Peptides rich in His provided the much needed buffering power and conductivity in the pH 6–7 gap.

Upon his return from Caltech, Harry hired a medical student, Olof Vesterberg, to help him devise a proper synthesis of the much-needed carrier ampholytes. After 3 years of slow progress, though, Svensson became Professor of Physical Chemistry at Gothenburg and the IEF team broke up. Vesterberg continued on this project in Stockholm and, in the spring of 1964, Svensson received a phone call from an excited Vesterberg, who appeared to have solved the problem. Well, he had been moonlighting and poring over textbooks of organic chemistry and had surfaced with a remarkable synthesis of the much wanted "carrier ampholytes": a chaotic synthesis, to be sure, as chaotic as a medical student could possibly devise. A most ingenious chaotic process, in fact, by which concoctions of oligoamines (from tetra- to hexa-amino groups) were reacted with limiting amounts of an α-β-unsaturated acid, acrylic acid.[6] Chaos generated order! In a steep voltage gradient, this army of synthetic amphoteres joined arms in an orderly fashion, with each assuming a (quasi) Gaussian distribution about its respective isoelectric point (pI) value.[7]

Was everything under control then? Well, with such a superb technique (one of the few able to counteract entropy's tendency to dissipate peaks via diffusion in the surrounding medium by producing sharply focused zones, no matter how carelessly the experiment was carried out), one would think yes, all was quiet on the western front!

Svensson and Vesterberg were indeed convinced that a breakthrough had been achieved, and they approached LKB Produkter AB, a Stockholm company, with a proposal for commercial production. As a safety measure, the two pilgrims were sent to the Mecca of Separation Science, Uppsala, to consult with the high priest of electrophoresis, Arne, *filius quondam Tiselii* (born Tiselius, in current slang). Poor Arne, who had been witnessing the steady erosion of his U-tube method, criticized them, infuriated at the notion that they would dare to bring forward a technique surely bound for failure, considering that macroions, on their approach to pI values, would likely aggregate and precipitate.

After such a discouraging pronouncement, the credit for faith and persistence goes to Herman Haglund (like Svensson, a former pupil of Tiselius), who was the head of a small team at LKB involved in separation techniques.[8] Even though he knew of the discouraging assessment of the "Maestro" on the Svensson–Vesterberg experiments, nevertheless, with the help of his colleagues Holmström and Davies, he decided to try to introduce into the market this novel, and quite revolutionary, methodology. He successfully squeezed from a reluctant LKB a bare 60,000 SKr (roughly 5000 euros in modern currency, a true pittance) and went into production of Svensson's vertical-density-gradient columns and of Versterberg's CAs. Thus, IEF was born as a preparative technique, requiring 110 and 440 mL columns for operation. An entire experiment, including column setup, focusing, elution, and the analysis of hundred of fractions, required a minimum of 1 week of hard labor! These columns and CAs were at first offered on a free-trial basis to the scientific community. Although during the 1960s the growth of IEF was painfully slow, by the beginning of the 1970s, especially due to the introduction of the analytical counterpart in polyacrylamide gels,[9] IEF enjoyed such a marked growth as to soon become a leading separation technique in all fields of biological sciences.

After such a long gestation period and hard trials, one would think that by then Svensson would have enjoyed a deserved glory and have been placed in the Separation Science Hall of Fame. No way! Early in 1970, "Svensson" figuratively died, only to be resurrected as "Rilbe." We all thought that this was a gentleman's gesture. Having remarried, we assumed that, to pay a tribute to the feminist movement, Harry had assumed his new wife's last name. Well, it was a "*nome de plume*," since of the powerful two-dimensional gel map analysis,[18] even today the most popular technique in proteome analysis. There is no way, though, to escape the crippling disease of aging on spaceship Earth. By the end of the 1970s, it was clear that a number of ailments was besieging Svensson's creature, namely uneven buffering capacity and conductivity, irreproducibility of CA synthesis, and the most-dreaded cathodic drift impeding proper focusing conditions, to name just a few.

It was with these problems in mind that we teamed up with LKB scientists to work out a totally new concept, immobilized pH gradients, that seemed to solve all of those problems at once.[19,20] This new work of art was unveiled on April 22, 1982, at the electrophoresis meeting organized by Stathakos in Athens and acclaimed with standing ovations. At least that is what we had hoped. In reality, we presented these data to an almost empty room, since most of the delegates had never been to Athens before and opted to enjoy lovely spring weather on the Acropolis, on the Licabetto, on the Plaka, strolling just about around any corner of the capital except the Hellenic Academy of Science, where the meeting was held. The rest is present-day history and is dealt with in most of the chapters in this book, so that it would be a shame for me to continue on these reminiscences.

REFERENCES

1. Svensson, H. Electrophoresis by the Moving Boundary Method. A Theoretical and Experimental Study. Almqvist & Wiksells Bok., Stockholm, 1946.
2. Svensson, H. *Sci. Tools* 3:30–35, 1956.
3. Svensson, H. *Arch. Biochem. Biophys.*, (Suppl. 1):132–140, 1962.
4. Svensson, H. *Acta Chem. Scand.* 15:325–341, 1961.
5. Svensson, H. *Acta Chem. Scand.* 16:456–466, 1961.
6. Vesterberg, O. *Acta Chem. Scand.* 23:2653–2666, 1969.
7. Rilbe, H. *Ann. N. Y. Acad. Sci.* 209:11–22, 1973.
8. Haglund, H. In Isoelectric Focusing (Arbuthnott, J. P. and Beeley, J. A. Eds.), Butterworths, London, pp. 3–22, 1975.
9. Righetti, P. G. and Drysdale, J. W. *Biochim. Biophys. Acta.* 236:17–24, 1971.
10. Kolin, A. *J. Chem. Phys.* 22:1628–1629, 1954.
11. Kolin, A. *Proc. Natl. Acad. Sci. USA*, 41:101–110, 1955.
12. Kolin, A. In Electrofocusing and Isotachophoresis (Radola, B. J. and Graesslin, D. Eds.), de Gruyter, Berlin, pp. 3–34, 1977.
13. Kolin, A. In "Electrophoresis '82" (Stathakos, D. Ed.) de Gruyter, Berlin, pp. 3–48, 1983.
14. Kauman, W. G. *Clas. Sci. Acad. Roy. Belg.* 43:854–868, 1957.
15. Chrambach, A. and Nguyen, N. Y. In "Electrophoresis '78 (Catsimpooloas, N. Ed.), Elsevier, Amsterdam, pp. 3–18, 1978.
16. Cuono, C. B. and Chapo, G. A. *Electrophoresis* 3:65–74, 1982.
17. Rilbe, H. *pH and Buffer Theory – a New Approach*, pp. 31–36. Wiley, Chichester, 1996.
18. O'Farrell, P. H. *J. Biol. Chem.* 250:4007–4021, 1975.
19. Bjellqvist, B., Ek, K., Righetti, P. G., Gianazza, E., Görg, A. and Postel, W. In Electrophoresis '82 (Stathakos, D. Ed.), de Gruyter, Berlin, pp. 61–74, 1983.
20. Bjellqvist, B., Ek, K., Righetti, P. G., Gianazza, E., Görg, A., Westermeier, R. and Postel, W. *J. Biochem. Biophys. Methods* 6:317–339, 1982.

I OVERVIEW

DAVID E. GARFIN[a] AND SATINDER AHUJA[b]

[a]*Biotechnology Consultant, 112 Kenyon Avenue, Kensington, CA 94708*
[b]*Ahuja Consulting, 1061 Rutledge Court, NW, Calabash, NC 28467*

Isoelectric focusing (IEF) is one of the most commonly used techniques for the separation of proteins.[1] IEF separations are based on the pH dependence of the electrophoretic mobilities of the protein molecules. Isoelectric focusing, as the name suggests, makes use of electrical charge properties of molecules to focus them in defined zones in a separation medium. It is the focusing mechanism that distinguishes IEF from other separation processes and makes it unique among the separation methods. In most separation methods, diffusion and interaction with the medium act to disperse the bands of separated molecules. In sharp contrast, the basic separation mechanism of IEF imposes forces on the molecules that directly counteract the dispersive effects of diffusion. During the separation process, the molecules in the sample accumulate in specific and predictable locations in the medium, regardless of their initial distribution. This focusing mechanism also distinguishes IEF from various modes of

Handbook of Isoelectric Focusing and Proteomics
D. Garfin and S. Ahuja, editors.

electrophoresis. In the other modes of electrophoresis, the applied electrical field moves molecules through the separation media at fixed rates, whereas the applied field in IEF establishes and maintains steady-state distributions of sample molecules. It is important to note that IEF is a high-resolution method that is well suited for both analytical and preparative applications.

I. SEPARATIONS BY IEF

From the viewpoint of separation technology, IEF is very impressive in resolving capabilities, yet elegant in its operational simplicity. It is favored by separation scientists but used insufficiently by protein biochemists. This may be due to the erroneous impression that IEF is more difficult than it really is. In actual practice, IEF is easy to understand and to perform; however, a complete understanding of IEF requires a strong grasp of a number of physical chemistry principles, including acid–base titrations. Fortunately, almost all the "hard work" has been done by the developers of this technique. As a result, routine experimentation can be carried out with IEF by simply preparing the samples and media and performing the run.

As mentioned above, IEF is unique among separation methods. This technique is applicable mainly to the fractionation of amphoteric molecules such as proteins and peptides that can act as both acids and bases. IEF is used mainly to separate proteins for analysis or purification. It measures the isoelectric points (pI) of proteins and uses the unique pI values of proteins to purify them. The pI of any particular protein is defined as the specific pH at which it carries no net electrical charge. Both analytical and preparative versions of IEF have been developed over the years.

The basis for electrofocusing lies in the pH dependence of the charges on the constituent amino acid side chains, non-proteinaceous adducts, and prosthetic groups of proteins. By subjecting proteins to electrophoresis in pH gradients, they become focused into well-defined, sharp zones at pH values corresponding to their individual pI. Very subtle differences in the pI values of proteins can be detected with IEF. Proteins differing in pI by less than 0.01 pH unit are routinely resolved by IEF, and separation of proteins with pI as low as 0.001 pH unit apart has been achieved.

Again, it is the focusing mechanism that distinguishes IEF from other separation processes. In all other methods of separation, diffusion and interactions with the medium act to disperse the bands of separated molecules. By contrast, the forces imposed on proteins in IEF counteract the dispersive effects of diffusion. In particular, IEF is distinguished from other forms of electrophoresis in which molecules move through the separation medium at fixed rates and in one direction from the point of application as long as the electric field is applied. During IEF separation, on the other hand, proteins in the sample approach their steady-state positions from any point within the separation medium and stop moving when they reach their pI values. However, once the electric field is removed, the steady-state distribution of molecules collapses and the previously separated proteins diffuse and mix.

Proteomics has heightened the interest in IEF. Proteomics is an approach to protein biochemistry, which has now emerged as a mature discipline. It can be defined as the systematic, massively parallel analysis of multiple proteins that is particularly well suited to screening experiments. The basic concepts of proteomics grew from considerations of two-dimensional polyacrylamide gel electrophoresis (2-DE), where IEF in polyacrylamide gels provides the first-dimension separation in 2-DE, the keystone separation method of proteomics. Because IEF gels can be made to match seamlessly with the second-dimension sodium dodecyl sulfate–polyacrylamide electrophoresis (SDS–PAGE) gel, 2-DE is the only multidimensional separation technique in which the resolution obtained in the first dimension is not lost when proteins are transferred to the second separation medium. Literally, thousands of polypeptides, including charge and size isomers, can be resolved in a single 2-DE gel. The number of useful proteins obtainable with 2-DE are restricted mainly by limitations imposed by available detection methods.

It was recognized from the outset that IEF provides a simple and reliable means for identifying charge isomers of proteins through differences in their pI values. Coupling IEF and SDS–PAGE in 2-DE affords biochemists and cell biologists a means for visualizing both the charge and size isomers of proteins in a single-gel entity. This is particularly relevant to charge isomers of proteins through differences in their pI values. It is also important in the study of post-translational modifications and their roles in protein function. Thus, IEF is and will remain an integral part of proteomic research and of protein biochemistry, in general.

Books devoted exclusively to IEF are rare. The classic monograph on this subject was first published in 1983 by Pier Giorgio Righetti (see photo in Chapter 2); he has been the main source of information on IEF.[2–4] This book complements the earlier texts. It is also a useful addition to the series of books on *Separation Science and Technology,* which is complementary to *Handbook of Bioseparations*[1] and *Bioseparation of Proteins*[5] in the series.

In this book, authored by various experts in the field, each chapter presents a unique description of a particular aspect of IEF along with special insights and practical tips about equipment, reagents, and procedures from the specialists in each field. The book is meant to be an initial reference source and yet provides in-depth knowledge of IEF to those engaged in protein research. It is designed to give readers an appreciation of the underlying principles of IEF and its various forms, so that this powerful technique can best be used to help solve biological problems. For the most part, the chapters provide the working knowledge of mathematical relationships, computational methods, or synthetic organic chemistry. Moreover, the chapters are intentionally independent of each another so that they can function as stand-alone descriptions of particular topics. The highlights of each chapter are given below.

II. EVOLUTION AND DEVELOPMENT OF IEF

A number of the most prominent people in the history of electrophoresis have contributed to the evolution of IEF. Through a slight stretch of the imagination, Chapter 2 traces this rich history starting in ancient Greece. However, the first IEF experiments most likely date back to 1912, when Ikeda and Suzuki achieved a separation of amino acids from plant protein hydrolysates in a three-chambered electrolysis cell. They noticed that the amino acids tended to arrange themselves according to increasing isoelectric point (pI) between the anode and cathode. In doing so, the separated amino acids formed a pH gradient between the electrodes. In 1929, Williams and Waterman extended this work by designing a multichamber compartment, which gave better resolution by reducing diffusion and convective disturbances. Although variable field strengths between the electrodes prevented formation of stable pH gradients, by limiting practical applications, these and other uses of the principle of IEF provided useful separation of peptides and proteins.

The concept of IEF was initiated by Kolin in his short note on the use of pH and density gradients for creation of "isoelectric line spectra." The main idea was to "focus ions in a continuous pH gradient" that was stabilized by a sucrose density gradient. One of the major obstacles in these pioneering experiments was the lack of suitable ampholytes for development of smooth pH gradients that were also sufficiently stable to allow true equilibrium focusing. Fortunately, these problems were solved, largely through the efforts of Svensson-Rilbe and Vesterberg (see Righetti's reminiscences on page xvii and the photos in Chapter 2).

III. THEORY AND SIMULATION OF IEF

The classical theoretical presentation of IEF generally relied on differential equations that could be solved analytically. It provided a description of the principle features of IEF that were adequate for most purposes. The analyses tended to be descriptions of the steady state and did not provide details of the dynamics of focusing. With computer analyses, the assumptions made in the classical theory are not necessary and the dynamics of focusing are more thoroughly described. As can be seen in Chapter 3, numerical simulations of IEF incorporate the dissociation rates of the components into the transport equations, generalize the discussion to multicomponent systems, and are not restricted to the steady state. The models successfully predict concentration, pH, and conductivity profiles as functions of time. They also describe transient states in the formation of the steady-state distributions of ampholytes and proteins and allow mechanisms of instability to be studied. This chapter is highly mathematical and provides the rational basis for the separations observed in IEF. Interested readers willing to make an effort to follow the mathematics and the descriptions of the numerical methods will learn that the entire focusing process with all of its nuances is amenable to physical and chemical analyses.

IV. GENERATION OF pH GRADIENTS

IEF is a very useful practical technique where simple methods for establishing and maintaining pH gradients are available. The pH gradient is essential for this technique, and the nature of the pH gradient largely determines the quality and usefulness of the separation. The two most widely adopted methods for generating pH gradients make use of different types of synthetic buffering molecules (see Chapter 4):

(a) Carrier ampholytes are amphoteric electrolytes that carry both current and buffering capacity. They possess both acidic and basic functional groups and form pH gradients under the influences of electric fields.
(b) Synthetic buffer compounds containing reactive double bonds, called acrylamido buffers, can be incorporated into polyacrylamide gel matrices. When used in the correct proportions, acrylamido buffers generate immobilized pH gradients under the influences of electric fields.

Both methods have their advantages and disadvantages and are described from a historical perspective in this chapter.

V. SLAB GEL IEF

IEF is most commonly carried out in polyacrylamide slab gels. Chapter 5 presents the practical aspects of slab gel IEF, including the use of several other gels that can be utilized for IEF. It provides empirical and "how-to" information. Experimental protocols are included for this purpose. The chapter is designed to be useful for novices in IEF and at the same time provide experienced users many useful tips and insights.

VI. TWO-DIMENSIONAL GEL ELECTROPHORESIS (2-DE)

Two-dimensional gel electrophoresis is a bioanalytical technique that provides high-resolution protein separation by integrating two independent electrophoretic separation methods. The first dimension employs the charge-based technique of isoelectric focusing, while the second step consists of size-based separation using SDS–PAGE. As a technique with wide utility, this method remains unsurpassed to date in its capacity to resolve polypeptides (see Chapter 6). It is the orthogonal separation that provides such high resolving power.

The high resolving capacity of 2-DE makes it a very desirable technique for proteomic analyses that aim to study thousands of proteins in a given sample. This chapter explains how to prepare proteins separated in the IEF gel for transfer to the second dimension and provides details of the 2-DE analysis. The technique is ideal for qualitative cataloging of the different protein "species" of a biological sample, and it is particularly useful for separating post-translationally modified protein isoforms. Moreover, 2-DE is well suited for quantitative studies of fluxes in protein synthesis and protein abundance. Proteins purified by 2-DE are readily accessible for analytical characterization conducted by mass spectrometry (MS). With the increasing analytical sensitivity afforded by MS (low fmol) and the decoding of several genomes, many of the proteins visualized on 2-D gels can be identified.

VII. PRACTICES AND PITFALLS OF SAMPLE PREPARATION

The key to successful performance of IEF and, by extension, successful 2-DE lies in sample preparation (see Chapter 7). Since protein mixtures vary widely in their properties, no single sample preparation procedure is universally applicable for all samples. Classical sample preparation for IEF relies on non-ionic or zwitterionic reagents to disrupt protein complexes and denature proteins to ensure that the subsequent electrophoretic separations are carried out on polypeptide monomers. Since IEF separates proteins based on isoelectric point, the single most powerful solubilizing reagent, SDS, is not normally usable, as it strongly attaches to proteins and often causes anomalous focusing and horizontal streaking in IEF. To approximate the denaturing power of boiling SDS under reducing conditions, IEF practitioners have relied on various cocktails of chaotropes, surfactants, and reducing agents. Chaotropes (such as urea, most commonly used for IEF) disrupt the hydrogen bonding at the protein surface and cause partial unfolding. When hydrogen in water bonds to the chaotropes instead of protein, the folded protein is more likely to open up and expose the (hydrophobic) interior. After the hydrophobic interior of the protein is exposed, the solubility is often compromised in aqueous solution. Therefore, it is desirable to have at least one surfactant present in the IEF cocktail to help solubilize the hydrophobic residues that are exposed as a result of denaturation in chaotropes. Many important considerations in sample preparation are also discussed. The aim of the work presented in this chapter is routine creation of high-resolution, streak-free IEF and 2-DE gels. This entails not only means for solubilizing proteins and keeping them soluble during IEF, but in the case of 2-DE, facilitates transfer of proteins from the first gel to the second gel.

VIII. PROTEIN DETECTION AND IMAGING

Gel-based IEF has a strong worldwide user base that supports a commercial pipeline of instrumentation as well as consumable products and will thus certainly remain a relatively low-cost, routine laboratory technique in the coming years. Detection methods used for visualizing proteins in IEF gels are described in Chapter 8. These techniques involve the use of many of the same dyes and stains that were developed for polyacrylamide electrophoresis gels but have been adapted for use with IEF gels. Modifications to accustomed procedures are necessary because most of the stains interact slightly with carrier ampholytes in IEF gels and to a lesser extent with the amine and carboxyl functionalities in the gel matrix of immobilized pH gradients. The important topics of image acquisition and analysis are also covered in this chapter.

IX. CAPILLARY IEF

Capillary IEF couples the automation capabilities of instrumental techniques with the high resolving capabilities of IEF. IEF in the capillary format (see Chapter 9) provides the high resolving power of conventional gel IEF and the automation capabilities of instrumental techniques such as capillary electrophoresis (CE) and high performance liquid chromatography (HPLC). The principle of capillary isoelectric focusing (cIEF) is similar to that of gel IEF: proteins migrate within a stable pH gradient formed by carrier ampholytes under the influence of an electric field. Upon attainment of equilibrium, proteins become focused within the pH gradient at the points where they have zero net charge, i.e., their isoelectric points (pI). Any diffusion of the focused protein away from its isoelectric zone will result in acquisition of charge, resulting in back-migration to the zone. The use of a narrow-bore fused silica capillary as the separation chamber provides efficient dissipation of Joule heat, enabling the use of very high electric fields (typically several hundred to a thousand V/cm). This allows separations to be performed in free solution, without the requirement for a gel as an anticonvective medium. The application of high field strengths provides high resolution (typically 0.02 pI units) and rapid analysis times. All steps in the analysis, including introduction of sample and ampholytes, focusing, and protein detection, can be performed automatically under instrument control, and the capillary can be reused for several hundred analyses. The ability to automate the IEF process and obtain quantitative information on resolved proteins is a driving force for the replacement of gel IEF by cIEF, particularly in industrial settings. All steps in the analysis can be carried out under instrument control, enabling the high-throughput applications desirable in industrial settings. Various relevant techniques of capillary IEF, along with discussions of their advantages and disadvantages, are presented in this chapter. Several pertinent applications of capillary IEF are also discussed.

X. PREPARATIVE IEF

IEF is well suited to preparative applications, and several different embodiments of preparative IEF devices have been developed. Methods for preparative IEF are described in several chapters. Preparative techniques range from simple slurries of gel-chromatography beads to specialized electrophoresis chambers. The pH gradients in preparative instruments are generated either by carrier ampholytes or by specialized applications of acrylamido buffer technology. Recovery of separated proteins from gel slurries is done by scraping out selected bands from the slurries, whereas recovery of proteins from the chambers is through access ports. Proteins obtained through preparative IEF can be quite pure and available for various other purposes. However, preparative IEF is most often used as part of a purification scheme, often upstream of other techniques. Nevertheless, as a consequence of the high sensitivity of analytical instruments such as mass spectrometers, applications of capillary IEF, which generates low quantities of proteins, can in some sense be considered preparative. Preparative IEF has also gained interest in the preliminary fractionation (the so-called prefractionation) of protein mixtures prior to 2-DE, both to decrease the complexity of the mixtures and to increase the accessibility of low abundance proteins.

A. Free-Flow IEF

Free-flow electrophoresis (FFE), also known as continuous free-film electrophoresis or continuous-flow electrophoresis (CFE), is one of the most versatile preparative-scale fractionation and separation techniques available to scientists of various disciplines (see Chapter 10). FFE utilizes a thin film of separation buffer that flows continuously in a laminar fashion between two closely spaced plates and an electric field that is applied perpendicular to the flow. FFE results in a differential deflection of charged sample components as they move toward the collection ports. This allows the high-throughput separation of all types of samples, such as low-molecular-weight organic compounds, peptides, proteins, protein complexes, membranes, organelles, and whole cells. FFE supports all modes of electrophoresis, such as zone electrophoresis (ZE), field-step electrophoresis (FSE), isotachophoresis (ITP), and IEF. In this chapter, the focus is on FF-IEF; it provides the reader with a comprehensive overview of its principles—covering all relevant parameters, the historical development, state of the art, and future instrumentation as well as the most recent applications. Wherever appropriate, general information about FFE has been included for completeness. Furthermore, the chapter touches on related technologies, such as multicompartment electrolyzers (MCE), to allow their proper differentiation.

Preparative separations are further discussed in Chapters 12 under the heading of Chromatofocusing.

XI. IEF AND PROTEOMICS

While proteomics has many broad definitions, the simplest may refer to scale (see Chapter 11). For the benefit of this chapter, proteomics has been defined as an ability to conduct high-throughput biochemistry or protein chemistry on a scale comparable with that achieved by molecular biology and its high-throughput counterpart, genomics. The advent of proteomics in the mid-1990s was made possible through a number of technical advancements for separating and identifying proteins, not the least of which was the sensitivity and automation capability of various MS technologies. However, all of the original techniques for separation of single unique proteins among the complex mixtures relied on IEF as a preliminary step, especially in 2-DE applications. In the 10-year period since the coining of the term proteome, IEF has become more than simply a tool utilized in 2-DE; in fact, it is now a recognized linchpin in the proteomics process. Researchers have begun to understand the value in high-resolution prefractionation steps to examine fractions with particular qualities, such as subcellular localization, prior to either 2-DE or 2-D liquid chromatography (2-DLC) for protein and peptide separation. IEF plays a central role prior to either of these applications, for example, in the separation and purification of organelles, the enrichment of high and low molecular mass, alkaline or hydrophobic proteins, or one of the several prefractionating devices used to enrich for proteins within a given pH range and compatible with micro-range (single pH unit) 2-D gels. This chapter deals with the proteomics aspects of IEF, with a focus on its role in prefractionation of biological samples and as the preliminary step in 2-DE, and an emphasis on reviewing the most recent developments.

It cannot be overemphasized that IEF remains an integral part of proteome analysis. It is central to the most commonly performed protein separation technique utilized in proteomics (2-DE), and in recent times it has become essential for protein prefractionation prior to 2-DE or other separation techniques. As such, it is a crucial element in allowing proteomics to access more than just the most abundant and readily solubilized proteins, and hence is a prerequisite for proteomic deep drilling of even the simplest of organisms. The role of IEF in 2-D gel-based proteomics is addressed in this chapter. Discussion covers the pros and cons of 2-DE proteomics, including the well-known underrepresentation of hydrophobic or membrane-associated proteins, highly alkaline proteins, high- and low-molecular weight proteins, and lower abundance proteins in relation to high abundance "housekeeping" proteins. Several methods based on IEF that are designed to overcome these problems are discussed in this chapter, including preliminary fractionation by IEF and modifications to the standard first-dimension IEF procedures.

The final two chapters diverge from description of purely IEF methods in order to highlight related, alternative approaches.

XII. CHROMATOFOCUSING

Chromatofocusing is a technique developed by Sluyterman and coworkers in 1978, predated by closely related work in ampholyte-displacement chromatography (see Chapter 12). The technique employs ion-exchange chromatography using a pH gradient (usually linear) to separate biomolecules with acid–base functionalities. It is principally used in the analysis and purification of proteins. Chromatofocusing was developed with the hope of making it a liquid chromatographic version of IEF, which performs separation and characterization based on the pI values of the protein. A third feature of IEF is the technique's high resolution, stemming from its ability to focus protein bands. While chromatofocusing generally separates proteins based on pI and focuses the protein bands better than salt gradient ion-exchange chromatography techniques, it does not realize the capabilities of IEF in terms of accurately determining pI or in achieving the same resolution. However, this chromatographic technique of pI separations is very useful as an analytical and preparative tool in the analysis and purification of proteins.

Although chromatofocusing is a chromatographic technique rather than an electrophoretic one, at the very basic level its similarities to IEF are evident. It is a chromatographic form of pI separation. To understand chromatofocusing, it is necessary to reorient the thought processes from electrophoretic to chromatographic concepts and terminology. As in IEF, the separation of proteins in chromatofocusing is based on the pI values of proteins. Differences in pI values are exploited to elute and recover proteins from complex mixtures. Development of pH gradients in chromatofocusing is not a steady-state process as in IEF, but a dynamic process in which pH is continually changing in the column. Moreover, there is only one direction of flow in chromatofocusing, from the top to the bottom of the column. In chromatofocusing, proteins are first bound to ion-exchange columns, either anion-exchanger columns or cation exchangers, and then they are serially eluted by means of pH gradients generated internally or externally to the ion-exchange column. When the pH of the eluent reaches the pI of a particular protein, the protein becomes uncharged and dislodges from the ion-exchange matrix and moves into the mobile phase, eventually eluting from the bottom of the column. This chapter discusses mainly anion-exchange chromatofocusing; however but the concepts hold equally well in the cation-exchange mode. Chromatofocusing is more of a preparative method than an analytical one, since the pI values obtained are only approximate. As a result, it is most often used as part of a protein purification scheme. An added advantage of this method is that it can be automated.

XIII. ALTERNATE ELECTROFOCUSING METHODS

Separation scientists classify IEF as one of the set of techniques termed equilibrium gradient methods. Chapter 13 considers IEF in this context and describes a set of related equilibrium gradient methods. In contrast to IEF, the set of described alternative methods are in the conceptual or very early development phase. They point to some of the possible routes that separation technology can undergo in the search for efficient techniques for fractionating and purifying proteins. This chapter provides the necessary mathematical background to clarify the basic concepts. The verbal descriptions of the methods have also been provided to help the readers comprehend the techniques that are presented and to encourage them to explore these methods further.

Alternative electrofocusing methods (AFMs) differ from IEF in that they do not focus solutes at their isoelectric points (pI). The AFMs analyzed in this chapter are part of a subset of the equilibrium gradient methods described by Giddings and Dahlgren, which use an applied electric field or electric field gradient as at least one of the counterbalanced forces on a focused solute. A complete binary set of EGMs would consist of dozens of pairs of forces, some of which are mentioned by Giddings in the context of field-flow fractionation (FFF) as well as in myriad variations on isocratic and gradient-elution chromatographies, each paired against a force, for example, hydrodynamic. It is unlikely that all possible binary pairs have been discovered, much less exploited, at this time. It does not appear that any ternary EGMs have been reported in the literature to date, so this area may still be considered "immature" and therefore ready for further development. The objective of this chapter is to describe what is known about this emerging set of methods in the hope that "gadgeteers" and theorists will explore this frontier and in doing so will help scientists to develop new tools for systems biology.

XIV. SUMMARY

In general, the concepts and manipulations of IEF remain virtually as they were in the early stages of its development. Advances in IEF have been mainly along the lines of refinements in reagents, techniques, and numerical analyses. The foundation technology of IEF gel electrophoresis, rather than simply withering away, actually appears to be undergoing a rebirth of sorts in the miniaturized world of IEF chips and microfluidic devices. It is believed that excellent technical contributions in this book will help protein biochemists to fully appreciate and utilize the powerful technique, that is IEF, in the quest to understand proteomes and beyond.

REFERENCES

1. Ahuja, S. *Handbook of Bioseparations*, Academic, NY, 2000.
2. Righetti, P. G. *Isoelectric Focusing: Theory, Methodology, and Applications*, Elsevier, Amsterdam, 1983.
3. Righetti, P. G. *Immobilized pH Gradients: Theory and Methodology*, Elsevier, Amsterdam, 1990.
4. Righetti, P. G., Stoyanov, A. and Zhukov, M. *The Proteome Revisited: Theory and Practice of All Relevant Electrophoretic Steps*, Elsevier, Amsterdam, 2001.
5. Sadana, A. *Bioseparation of Proteins*, Academic, NY, 1998.

2 EVOLUTION AND DEVELOPMENT OF ISOELECTRIC FOCUSING

ÁKOS VÉGVÁRI AND FERENC KILÁR

Department of Analytical Chemistry, Faculty of Science and Institute of Bioanalysis, Faculty of Medicine, University of Pécs, Pécs, Hungary

I. INTRODUCTION

Isoelectric focusing (IEF) has had a long evolutionary process. The development of this technique was based on several original ideas and needed tremendous work. This chapter summarizes the progression of events that led to successful preparative and analytical applications of isoelectric focusing from the earliest "electrophoretic knowledge" until the modern theoretical advances.

II. THE RISE OF ELECTROPHORESIS

A worthwhile prelude to the evolution of successful IEF is a tracing of the history or electrophoresis in general. "Electrophoresis" has a long history, which probably starts as early as 600 B.C. in ancient Greece.

Although there is no written record from that time, we know, as A. Kolin—a pioneer of the technique—pointed out in his lecture about "Evolution of Ideas in Electrophoretic Developments",[1] that Thales of Miletus—generally considered to be the father of Greek science—had known the special properties of amber ("electron" in ancient Greek). When amber is rubbed, it attracts small light objects, such as pieces of straw and then it repels them after contact. Thales might also have observed the attraction of oppositely charged dust particles toward and repulsion of particles of like charge from an electrified piece of amber by positioning the amber in a light beam in a dark room and viewing it sideways. The phenomenon can be considered as electrophoresis in *air*! Unfortunately, the ancient knowledge disappeared from Europe (and rest of the world) for many centuries, and electrophoresis in *liquids* was discovered only after the invention of the Voltaic battery in 1800[2] that provided an electrochemical energy source.

A mere 9 years after Volta's historic publication, Reuss, a professor in physics at the University of Moscow, who was studying electrical conductivity of moist soil in a garden on the shores of the Moskva River, discovered electrophoresis.[3] This was at a time when the nature of electric current was mysterious—people spoke of an "electric effluvium." Ohm's law was discovered 28 years later and Faraday's laws of electrolysis after 32 years. In these early days, the intensity of electric currents was semiquantitatively estimated by observing the rate of evolution of bubbles in an acidic solution. Physicists' attention eventually turned to electrical conduction and they intensively investigated all kinds of conductive medium. Reuss' electric battery was a Voltaic pile consisting of a multiple-layer sandwich of 92 silver rubel coins separated from 92 zinc discs by layers of acidic-soaked cloth. Reuss was surprised that the distance between his electrodes in the soil did not significantly affect the current. He set up a simple apparatus in the laboratory and repeated his experiments under better controlled conditions. Instead of soil the test apparatus contained a layer of quartz sand. In his first experiment Reuss discovered electroosmosis as a rise of liquid in the positive electrode leg. There was no effect in the absence of the sand. He reported it on April 15, 1809 in a lecture entitled "On a New Effect of Galvanic Electricity."

In the second experiment, Reuss replaced the sand with clay particles and inserted two vertical glass tubes into a slab of wet clay (Figure 1). He thus discovered electrophoresis by observing that clay particles migrated upward in the tube filled with conductive solution at the positive pole.

Physicists (but not yet analytical chemists) continued the investigations some decades later by developing the theory of electrophoresis. Helmholtz' (the co-discoverer of the law of conservation of energy) famous mathematical formulation of electrophoresis[4] was followed by a common theory for electrophoresis and electroendosmosis based on the notion of an interfacial electric double layer.[5] This idea formed the basis

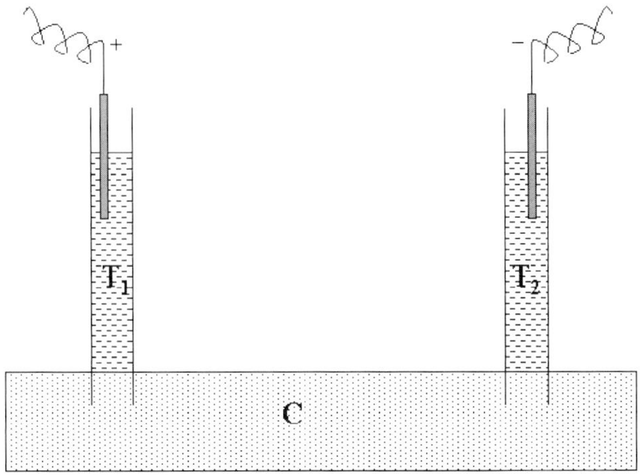

FIGURE 1 Reuss' experimental setup of electrophoresis. C is the moist clay slab; T_1 and T_2 are glass tubes.

of a more sophisticated theory of the diffuse ionic atmosphere in the subsequent treatments of Gouy in 1910, Chapman in 1913, and Stern (a Nobel Laureate in physics "for his contribution to the development of the molecular ray method and his discovery of the magnetic moment of the proton" in 1943) in 1924.[6–8] The theories originally designed for plane surfaces were extended to spherical configurations by the introduction of the theory of counter-ion atmosphere and ionic migration by Debye (a Nobel Laureate in chemistry "for his contributions to our knowledge of molecular structure through his investigations on dipole moments and on the diffraction of X-rays and electrons in gases" in 1936) and Hückel in 1923.[9] We also have to mention Arrhenius' (a Nobel Laureate in chemistry "in recognition of the extraordinary services he has rendered to the advancement of chemistry by his electrolytic theory of dissociation" in 1903) revolutionary theory of ionic dissociation, which assumed that supposedly immutable atoms radically change their chemical properties by acquisition of an electric charge in becoming ions.[10]

Experimental work complemented the theoretical developments. Kohlrausch's—a notable pioneer in the study of electrolyte conductance—discovery of the laws for concentration shifts and boundary migration in electrolyte columns provided a foundation for some of the most important recent developments in electrophoretic methodology.[11] Lodge, in 1886, proposed the moving-boundary method and zone electrophoresis in gels in his studies on direct observation and measurement of migration rates of ions in solution.[12] Tiselius (Figure 2)—a Nobel Laureate in chemistry "for his research on electrophoresis and adsorption analysis, especially

FIGURE 2 **Arne Tiselius.**

for his discoveries concerning the complex nature of the serum proteins" in 1948—has perfected the former electrophoretic method for analytical use in 1930.[13]

Early experiments were carried out on hemoglobin,[14] but the potential of electrophoresis as a tool in biochemical investigations was not realized before Tiselius developed his method. With the advent of Tiselius' apparatus (Figure 3), boundaries formed by electrophoretically migrating proteins in buffer solutions could be recorded optically by measurements of light absorption or refractive index. This analytical device, in which the sample ions are subjected to electrophoresis in free buffer (i.e., without an anticonvection medium), was supplemented by the introduction of paper electrophoresis in the late 1940s.[15]

The advantages of simplicity in technique and completeness of zonal separations on filter paper were gained at the expense of the emergence

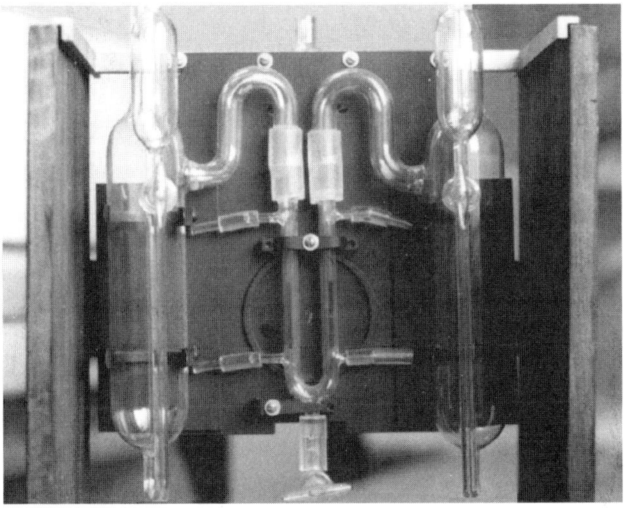

FIGURE 3 **The electrophoresis apparatus from 1930 invented by Arne Tiselius.**

of new artefacts. Pronounced electroosmosis and streaming caused by evaporation of water as well as non-uniformity of buffer concentration and temperature along the paper strip, made filter-paper electrophoresis unreliable means for measurement of electrophoretic mobilities for the characterization of compounds. Adsorptive interaction between filter paper and some compounds caused not only retardation and possibly immobilization of some zones, but also "tailing" which broadened zones and degraded the separation pattern. This led to replacement of paper by other materials, such as cellulose acetate in which some of these problems are less severe. The use of a thin supporting matrix provides, however, one major advantage over filter-paper methods: the possibility of cooling the strip. This permits high degrees of separation by means of "high-voltage electrophoresis."[16]

The search for materials, in which the above-mentioned artefacts were less pronounced than in paper or in thin plastic films, led to the discovery of many different types of gels with suitable properties, most notably polyacrylamide gel (originated by Lodge[12]).

The transition from a paper- or thin-film matrix to a gel had an important advantage of adding a significant third dimension, which permitted an increase in the sample quantity for preparative separations. But the gel layer can also be made very thin so as to permit cooling and application of high potential gradients to achieve rapid separations. There is one property of gels that can be utilized concurrently with electrophoresis, which makes them superior to other matrices: the possibility of superimposing a molecular sieving effect upon electrophoresis, thus adding the capability of resolution on the basis of differences in molecular dimensions. Smithies was the first to introduce such a medium,[17] a starch gel, which, unfortunately, had a low but not negligible content of charged groups which caused some adsorption of proteins and gave rise to an electroendosmotic flow. It was also difficult to manufacture starch batches with reproducible properties. The starch gel was soon replaced by polyacrylamide gels, independently by Raymond and Weintraub,[18] Davis and Orstein,[19] and Hjertén.[20] With polyacrylamide as a nearly ideal anticonvection medium, the resolving power of electrophoresis was increased, diffusion was reduced, and migration of the sample components in sharp zones was enabled. Hjertén showed that gels of neutral agarose[21] are superior to charged agar gels (which contain sulfate groups) for electrophoresis and that they possess much larger pores than polyacrylamide gels.

III. KOLIN'S "ISOELECTRIC SPECTRA"—THE ARTIFICIAL pH GRADIENT

Probably the first IEF experiments date to 1912 when Ikeda and Suzuki[22] achieved a separation of amino acids from plant protein hydrolysates in a three-chambered electrolysis cell. They noticed that the amino acids tended to arrange themselves according to increasing isoelectric point (pI) between the anode and cathode. In doing so, the separated amino acids formed a pH gradient between the electrodes. In 1929, Williams and Waterman[23] extended this work by designing a multichamber compartment, which gave better resolution by reducing diffusion and convective disturbances. Although variable field strengths between the electrodes prevented the formation of stable pH gradients, limiting practical applications, these and other uses of the principle of IEF provided useful separation of peptides and proteins.[24]

One of the major obstacles in these pioneer experiments was the lack of suitable ampholytes for the development of smooth pH gradients that were also sufficiently stable to allow true equilibrium focusing. Fortunately, these problems were solved, largely through the efforts of three chemists: Kolin, Svensson (later called Rilbe), and Vesterberg.

IEF, as a concept, was initiated by Kolin (Figure 4) with his short note on the use of pH and density gradients for creation of "isoelectric line spectra." This idea of "focusing ions in a continuous pH gradient," stabilized by a sucrose density gradient, was presented in 1954.[25]

Kolin described the separation and concentration of proteins by electrical transport in a pH gradient. The components of a given mixture are separated from each other simultaneously and are sorted in a spatial arrangement, hence referred as an "electrophoretic spectrum." The term "spectrum" was used to designate a sorting in the same sense in which one speaks of "mass spectra" or "frequency spectra." Kolin considered two kinds of electrophoretic line spectra: (i) "mobility spectra" and

FIGURE 4 Alexander Kolin.

(ii) "isoelectric spectra." In the first kind of spectrum, the components are separated according to their differences in electrophoretic mobility, whereas in the second kind of spectrum the components are separated according to differences in their isoelectric points. Naturally, both types of spectra may be occasionally superimposed upon each other (Kolin suggested the use of the term "electrophoretic line spectra" in such cases). Kolin placed the substances to be separated at the interface between an acidic and a basic buffer in a Tiselius-like apparatus and after an appropriate diffusion time, applied an electric field. This generated a pH gradient, later to be called an "artificial" pH gradient by Svensson upon his invention of "natural" pH gradients (see below). Kolin was able to obtain "isoelectric line spectra" of dyes, proteins, cells, microorganisms, and viruses on a time scale ranging from 40 s up to a few minutes; rapidity still unmatched in the field of electrophoresis. Concomitant with the pH gradient, a density gradient, an electrical conductivity gradient and a vertical temperature gradient were acting upon Kolin's separation cell.

Kolin's initial paper was followed by a long series of papers showing his quite prolific innovative ability.[26–28,1] Kolin pointed out the importance of using electrolytes with high buffering capacity and stabilizing the pH gradient against convective mixing.[25,26] His approach consisted of two steps: (i) the generation of a pH gradient along the separation path by diffusion between two limiting solutions, which were titrated to the starting and ending pH values, followed by (ii) a quick electrokinetic separation. The substances to be separated were placed at the interface of an acidic and a basic buffer, which were allowed to diffuse against one another while subjected to an electric field. The separation force in these experiments was, therefore, the resultant of several factors: a pH gradient, a density gradient, an electrical conductivity gradient, and a vertical temperature gradient. Unfortunately, the pH gradients were unstable because of the rapid migration of the buffers during electrolysis, and the separated components could not be easily recovered.

Kolin described the three related effects of IEF during the course of separation that play important roles: (i) "Isoelectric condensation", in which amphoteric molecules are swept toward their isoelectric zones where their charges and accumulate. (ii) "Isoelectric evacuation", wherein oppositely charged, non-isoelectric ions located on either side of the isoelectric zone move away from this zone, to create a "protein ion vacuum" on each side of it. Protein molecules removed from the isoelectric zone are swept away by the electric field. (iii) The isoelectric condensation zones must exhibit electrochemical stability. Molecules removed mechanically or thermally from such a zone acquire a charge in the pH gradient and, thus forced to be returned to the isoelectric zone.

A similar principle, called electrophoretic focusing of ions (EFI) (focusing ion exchange) was reported by Friedli and Schumacher[29] for separation of rare-earth mixtures in combined proton (pH) and ligand (pL) gradients. EFI spectra of La–Tb and Eu–Lu groups were obtained in the course of 5 min. However, not exactly focusing, but electrophoresis perpendicular to a pre-established pH gradient was practiced already in 1952 by Michl[30]: it was an embryonic form of "titration (pH/mobility) curves," as later developed utilizing IEF in the first dimension.[31] pH gradients generated by diffusion were also used by Stahl[32,33] for obtaining "titration curves" in thin-layer chromatography (pH/R_f curves) and later by Tate[34] in paper electrophoresis (determination of ionization constants of nucleotides); they were also used by Jokl et al.[35] in hydro-organic solvents. Jokl observed that the "isoelectric lines" of proteins were much sharper and more stable than those of non-proteins—a general attractive property of IEF. However, he assumed that this sharpness is due to the stable equilibrium of protein molecules in the isoelectric plane. He then clearly explained what happens when a molecule (ampholyte) gains charge by migrating from the focused zone in the electric field (i.e., the "isoelectric evacuation" effect).

Early analytical biochemists designed an "electrodialysis apparatus" with compartments separated by membranes and containing buffers of different pH values.[23,36] The apparatus consisted of 14 compartments filled with ampholytes in solutions of different pH values separated by parchment membranes. After 60 h, the highest concentration of individual species of ampholytes was found in compartments having pH values nearest to their respective isoelectric points. Some later modifications of this apparatus used separate containers connected in series by electrolyte bridges. This method may be considered as an intermediate step between standard electrophoretic technique and the method of "isoelectric spectra." But, in spite of the time lapse since 1929, no successful attempts were made to turn from a pH step-function to a continuous pH gradient.

Interestingly, Kolin also described in his first paper how to concentrate components of a broad starting zone by the difference between the conductivities in the zone and the following electrolyte (when the conductivity is higher in the latter one). This phenomenon is frequently utilized today in capillary and microchip electrophoresis experiments when analyses of very low concentrations of proteins require pre-concentration in order to detect them. It is difficult to understand as to why no instrument factory has taken up any of Kolin's ideas and constructions for separation and analysis of macromolecules and cells. However, the International Electrophoresis Society honored him (and Rilbe) with its first "Award for Outstanding Contributions to the Field" in the year 1981.

IV. SVENSSON'S IEF—VESTERBERG'S SYNTHESIS

The pH gradients used by Kolin were very short and unstable with time. This strengthened the desire of Svensson (Figure 5) to develop a new IEF method. In 1956 he published a paper on the concept of transport numbers of ampholytes.[37] The achievement of pH gradients of adequate stability and durability was a decisive contribution of Svensson and Vesterberg.[38–41]

Svensson (who in 1968 changed his name to Rilbe) published the basic theory of IEF[38] and, in a later paper,[39] experimentally verified that pure ampholytes had conductivities in agreement with theoretical predictions. Isoelectric ampholytes with zero net charge were found to have appreciable conductance if the pK values of the dissociating groups were close. If many such ampholytes, isoelectric at various pH values, were available, then on electrolysis they would be able to dictate, from the anode to the cathode, a smooth pH course without any deep conductivity minima. This was the very basic concept of IEF, and the desirable ampholytes were later called *carrier ampholytes* (short for "amphoteric electrolytes").

The separation of haemoglobins was also described.[40] However, the lack of sufficient number of suitable ampholyte molecules delayed the

FIGURE 5 **Harry Svensson.**

process of development. The real breakthrough for the scientific world arrived when Vesterberg (Figure 6), a medical student at the Karolinska Institute, joined Svensson's research group and started to work on synthesis of new ampholytes.

Righetti begins his excellent monograph[42] by writing in the preface about the difficulties Svensson had to face: not only scientific problems but also human frailties. As Righetti narrates , the "story" of ampholytes goes as follows: "In 1959 Svensson was on a visit to Pauling in California Institute of Technology. He was already outlining the basic theory of IEF but was frustrated to find that even in the United States, the chemical catalogues listed no suitable ampholytes. After his return to Sweden, as a freelance at the Karolinska Institute (note that professorships were just as hard to come by then as now), Svensson teamed with Vesterberg (a medical student) to solve the practical aspects of the problem. After 3 years of

FIGURE 6 Olof Alfred Yngve Vesterberg.

slow progress, Svensson became a Professor at Gothenburg and the IEF team broke up. Vesterberg continued with the problem in Stockholm and, in the spring of 1964, Svensson received a call from an excited Vesterberg, who appeared to have devised a satisfactory synthesis of carrier ampholytes. After careful checking of the data, Svensson and Vesterberg were convinced that a breakthrough had been achieved and they approached LKB Produkter with a proposal for commercial production. At that time, Uppsala was the Mecca of separation techniques, hence the two pilgrims from LKB were sent to consult the high priest of electrophoresis, Professor Tiselius. They were disconcerted to be told that an electrophoretic separation technique in which macromolecules would be driven to their isoelectric points could never work as the macroions would aggregate and precipitate. After such a discouraging verdict the credit for faith and persistence goes to H. Haglund (a former pupil of Tiselius like Svensson), who was head of a small team at LKB involved in separation techniques. Knowing of the Svensson–Vesterberg experiments and the discouraging assessment of the 'master', Haglund with his colleagues Holmström and Davies decided to try to market the 'gimmick.' They successfully squeezed from a reluctant LKB a bare 60,000 SKr, and went into production making Svensson's vertical density gradient columns and the first batches of Vesterberg's carrier ampholytes. These were offered on a free trial basis to the scientific community."

Vesterberg's elegant approach to the synthesis of carrier ampholytes[41,43] has enabled today's IEF. He filed a Swedish patent in August 26, 1964, which was extended in the English version (U.S. patent no. 3,485,736; December 23, 1969).

In his summarizing paper about the history of electrophoretic methods,[44] Vesterberg himself recalls the struggle in seeking indefatigably for useful carrier ampholytes, his synthetic approaches that at first were all failures and, finally, the "Eureka" solution: "At the Karolinska Institute, Svensson formed a research group, which I had the opportunity to join. The search to find suitable carrier ampholytes had high priority. In spite of scrutinizing catalogues of commercially available chemicals, only a few suitable materials could be found. It was especially difficult to find carrier ampholytes with isoelectric points in the pH range 4–8. Nor did we receive any help from organic chemists; a typical reply was, 'to get the many substances needed would constitute a huge task.' By partial hydrolysis of haemoglobin or whole blood we kept producing oligopeptides that were used as carrier ampholytes. This "bucket-scale" work, a laborious task, delayed us from extending IEF to new fields, e.g., focusing in gels. Despite treatment with carbon and other sorbents, we could not obtain colorless peptide preparations. However, IEF showed that the color could be focused in a few discrete zones. In 1963 Svensson left for a Professorship in Gothenburg and I obtained some laboratory space at the Karolinska Institute in Professor Theorell's laboratories at the Nobel Medical Institute. It was very annoying that the peptides that I used as carrier ampholytes gave zones of similar color to myoglobins, which I studied. This promoted a search for uncolored synthetic ampholytes with suitable buffer capacity and conductance. From my chemistry studies I had memorized the fact that owing to mutual influences within a molecule, identical protolytic groups in polyvalent acids and bases may have widely different p*K* values. I made an extensive study of possible synthetic methods for amino acids. At first I was disappointed because the methods found were unsuitable for some reason. Finally, in 1964, I tried to attach carboxylic acids to amines. By boiling under reflux a mixture of acrylic acid and polyvalent amines I obtained ampholytes with many protolytic groups having suitable p*K* values and isoelectric points. The first syntheses were encouraging. I worked very hard and developed modified recipes giving improved properties of the carrier ampholytes, which finally fulfilled all the desired criteria." His historical notes on the results are shown in Figure 7.

Vesterberg's synthetic procedure is basically as follows a mixture of oligoamines (the more heterogeneous the better, e.g., triethylene triamine, tetraethylene pentamine, pentaethylene hexamine) is reacted with an α-β unsaturated compound (the best being acrylic acid) to form a highly complex mixture of aliphatic oligo-amino and oligo-carboxylic acids. These have pI values in the pH range 3–10, with small (pI–pH proximal) values and are thus able to buffer and conduct in this pH

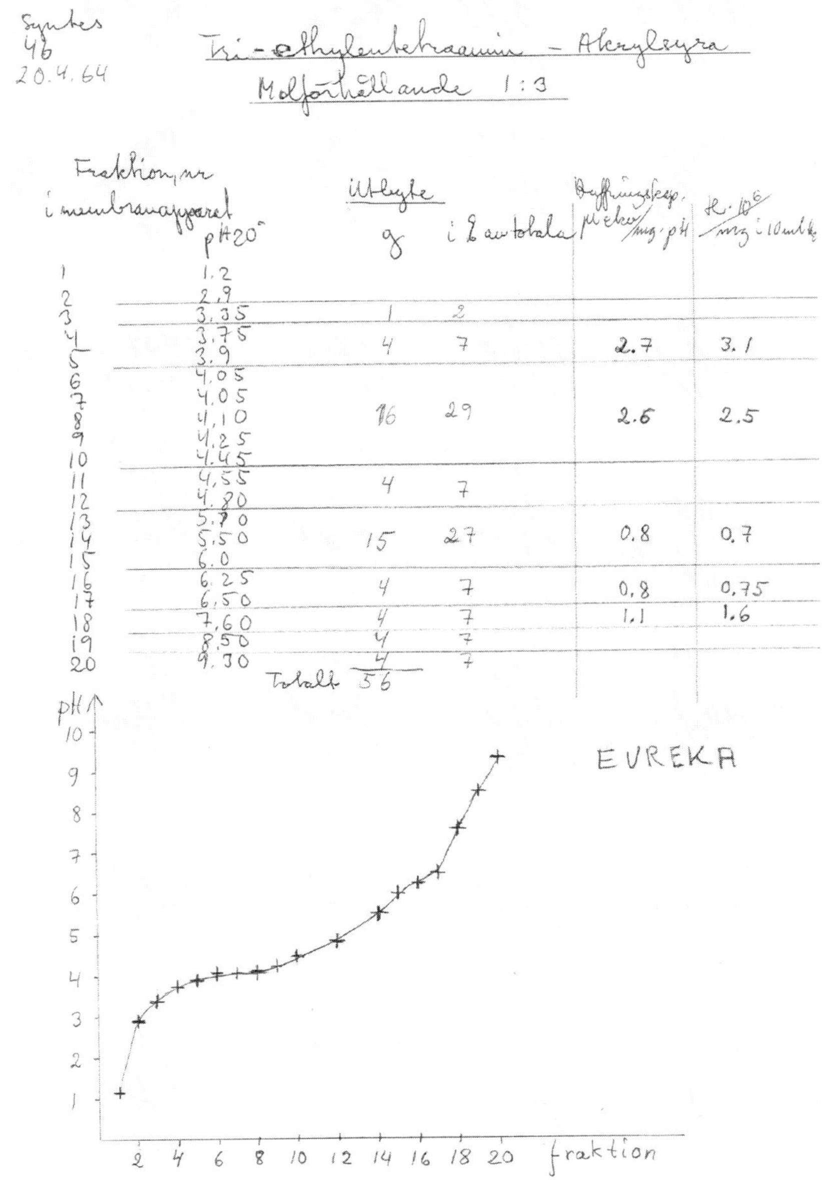

FIGURE 7 "Eureka," in the laboratory notes of Vesterberg when the synthesis successfully created a pH gradient of ampholytes.

region. Although, it was never clearly stated in the literature, it is obvious that LKB's commercial Ampholines are indeed the carrier ampholytes first devised and patented by Vesterberg. According to the patent, they are "polyprotic amino carboxylic acids, containing at least four weak protolytic groups, at least one being a carboxyl group and one a basic nitrogen atom, but no peptide bonds".

The advantages of these carrier ampholytes were obvious: upon being focused in density gradients of sucrose, two myoglobins could be separated although the difference in their isoelectric points (ΔpI) was only 0.05 pH unit. To resolve them, it was necessary to create a very shallow pH course. The resolving power was thus better than 0.05 pH unit, whereas a theoretical resolving power of 0.02 was calculated.[45] Vesterberg has also designed and built apparatus for fractionation of the synthetic products and also columns for IEF and separation of proteins in density gradients.[46]

V. PROGRESS IN PREPARATIVE AND ANALYTICAL IEF

One should notice that IEF was born as a preparative technique run in vertical glass tubes. Rilbe and his collaborators designed apparatus for rapid and convenient preparation and focusing in short density gradients.[47] Jonsson and Rilbe developed a method for IEF, which permitted the convenient spectrophotometric evaluation of the separation[48] that was continued by Fredriksson.[49] Procedures for the preparative purification of proteins have also been described[50] and focusing in granulated gels (such as Sephadex™) can also be mentioned.[51]

Over the years, several synthetic approaches for ampholytes generation have been described, even high molecular weight species for fractionation of peptides, but all have been based on the classical synthetic approach of Vesterberg. Serva introduced another synthesis (their ampholytes are called Servalyte™), which has been described by Pogacar and Jarecki[52] and by Grubhofer and Borja.[53] A new generation of buffering ampholytes was introduced by Pharmacia under the trade name of Pharmalyte™.[54,55] However, all these three follow the basic idea of Vesterberg: the synthesis of amphoteric compounds from polyamines and organic/inorganic acids in a way to obtain the most heterogeneous mixture of products. Over the years other laboratory synthesis methods saw the light of day as well.[56–60]

A radically different approach to the generation of pH gradients, which does not rely on carrier ampholytes or on prolonged electrolysis, is to take advantage of the temperature coefficient of the pH of a buffer as well as of the pI of the proteins to be separated.[61,62] A pH gradient can be established in a buffer solution within seconds by taking advantage of the temperature dependence of the p*K* value. By establishing a temperature gradient within the buffer, pH gradients can be obtained that span a pH range of about 1 pH unit.

Troitzki et al.[63] formed pH gradients by using common buffers in gradients of organic solvents, such as ethanol, dioxane, glycerol, or in polyol gradients such as mannitol, sucrose, and sorbitol. Taking advantage of the pH variations of these buffers in different concentrations of these solvents, they were able to generate gradients of approximately 1.5 pH units in different regions of the pH scale. These pH gradients were stable up to 12 days of IEF under voltages up to 1000 V.

Thousands of laboratories have been benefited from the use of horizontally run flat-bed gels for various types of electrophoresis, especially IEF, using equipment and apparatus developed by Vesterberg[64,65] and by others, such as Multiphor™ (since 1973) and Ultraphor™, which were originally marketed by LKB and later by Pharmacia. IEF in ultrathin polyacrylamide layers, as developed by Görg et al.,[66] represents one of the most interesting breakthroughs in polyacrylamide gel slab IEF.

The two major forms of analytical IEF are two-dimensional polyacrylamide gel electrophoresis and capillary IEF. Since both are described in separate chapters in this book only the developmental phases of the techniques are mentioned here. In the 1960s, Hjertén (Figure 8) developed the first

FIGURE 8 Stellan Hjertén.

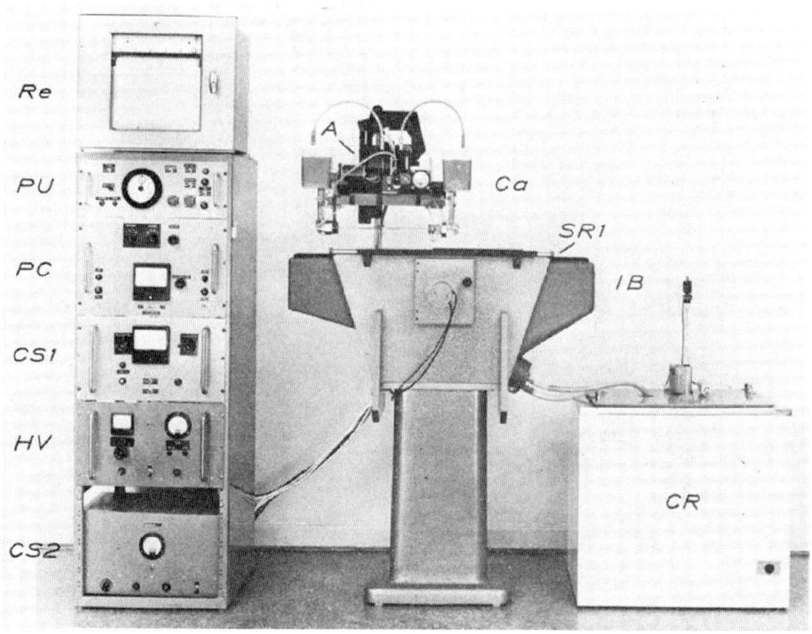

FIGURE 9 **The setup of the "Free zone electrophoresis" apparatus of Hjertén.**[67]

"capillary electrophoresis apparatus," the *Free-zone electrophoresis in a rotating tube* (Figure 9).[67] Lundahl and Hjertén[62] and Hjertén[68] modified this apparatus for use in IEF. Coating of the tube with methyl cellulose eliminates electroosmosis. The tube is scanned in UV light and the ratio of absorbances at 320 and 280 nm is recorded. At the two tube extremities, polyacrylamide beads are packed to avoid convective mixing of anolyte and catholyte with the Ampholine™ solution in the tube. A cellophane membrane at the tube ends prevents hydrodynamic liquid streaming.

VI. THE IMMOBILIZED pH GRADIENTS

The latest evolutionary step in the development of IEF was the immobilized pH gradients (IPG) containing buffering substances covalently bound to the gel matrix, which provides for indefinitely stable pH gradients[69]. Weak acrylamido derivatives containing either carboxyl groups or tertiary amino groups, supplemented by one strongly acidic and basic derivative, are available commercially under the trade name, Immobiline™, from Pharmacia Biotechnologies, Uppsala, Sweden. These monomers finally made it possible to create and maintain stable pH gradients in electric fields until the attainment of steady-state conditions.

FIGURE 10 Pier-Giorgio Righetti.

Immobilized pH gradients were originally invented in the small company Aminkemi, during in the synthesis of carrier ampholytes for LKB, and were first described in a Swedish patent by Gasparic et al.[70–72] Righetti (Figure 10) and his group were also active in similar developments, and together with Bjellqvist and a German group they published the basic scientific paper[69] on the subject. This technique proved to be capable of higher resolving power than carrier ampholytes; a pI difference of 0.001 pH unit being sufficient for electrophoretic separation. Altland et al.[73] and others later found that the simultaneous presence of free carrier ampholytes in immobilized pH gradients improves isoelectric separation. A comprehensive review on the use of immobilized pH gradients has been published by Righetti.[74]

Several problems, particularly related to the alkaline chemicals used to create and maintain the pH gradient, at first impeded the spread of this method worldwide. These problems could be grouped into three categories: (i) autopolymerization to form oligomers and *n*-mers; (ii) hydrolysis to free acrylic acid and a diamine; and (iii) formation of N-oxides due to persulfate oxidation during polymerization.[75] The second generation of these compounds (Immobiline II) was introduced in the summer of 1988,[76] offering a stable and reliable technique of unrivalled versatility and resolving power. The autopolymerization and hydrolysis of the alkaline Immobilines™ were fully eliminated with this version. However, other artefacts, such as severe smears were observed in gradients as wide as pH 4–9, encompassing neutrality and containing the pK 7.0 Immobiline™ as one of the buffering ions.[77] The effect was found to be directly proportional to the total amount of this pK 7.0 species present in the gradient formulation. The problem was traced back to the presence of oligomers in some commercial preparations, probably formed during the synthetic step, and has since been eliminated.

VII. TWO-DIMENSIONAL GEL ELECTROPHORESIS AND BLOTTING OF PROTEINS

Smithies and Poulik published the earliest application of electrophoresis in two dimensions in gels in 1956.[78] However, the starch gels they used were far from optimal. Margolis and Kendrick used polyacrylamide gel (PAG) preferably with a gradient.[79] Improvements were obtained by using IEF of proteins in a rod of polyacrylamide gel for separation according to charge in the first dimension, followed by electrophoresis perpendicularly in the second dimension in a polyacrylamide gel slab as described by Dale and Latner,[80] and Macko and Stegemann.[81] In 1970, Stegemann[82] introduced IEF in PAG followed by electrophoresis of proteins in a gel slab containing SDS to increase the negative net charge of proteins and to utilize the relationship between their size and mobility. High-resolution separation of proteins by two-dimensional gel electrophoresis was obtained after pretreatment of the samples in hot SDS–urea solutions and IEF followed by SDS–PAGE as described in 1975 by O'Farrell,[83] Klose,[84] and Scheele.[85] A few years later, Anderson and Anderson[86] described the high-resolution separation of human serum proteins and the semi-automated ISO-DALT system allowing the parallel use of 20 two-dimensional electrophoresis gels.

The transfer of proteins by electrophoresis from a gel slab on to a sheet of nitrocellulose or some other material is very helpful for the identification of protein spots or bands in gel slabs, after one- and two-dimensional electrophoretic separations.[87] After 1985, a few groups developed general methods for direct N-terminal[88] and internal sequencing[89] of gel separated proteins not only from l-D but also from 2-D matrices. A technique, called "press-blotting," was developed using gelatin-coated nitrocellulose membranes to perform sensitive quantitative immunodetection of peptides after gel IEF.[90]

One of the major difficulties in establishing 2-D maps in the early systems was the variability of spot position in the first (focusing) dimension due to both batch-to-batch variations of carrier ampholytes and pH gradient decay in conventional IEF. This rendered spot identification, pattern matching, and inter-laboratory comparison quite problematic. IPGs proved to be useful in 2-D applications, resulting in constant zone position and pattern consistency. Additionally, IPGs allowed the creation of reproducible, non-linear pH gradients prepared to account for spot distribution and frequency along the pH scale.[91,92]

VIII. CAPILLARY IEF

Capillary IEF is one of the separation techniques with the highest resolving power. After the first experiments performed by Hjertén and co-workers in the mid 1980s,[93–95] hundreds of papers have appeared about its methodological aspects and utilization. Hjertén—considered the "Father of capillary electrophoresis"—performed IEF in coated capillaries, and showed that the pattern can be obtained by hydrodynamic or electrokinetic mobilization.[96]

Further methodological development in capillary IEF included the use of uncoated capillaries[97,98] and the sequential injection protocol.[99] Recent years brought an extensive increase in the applications of this technique employing its exceptionally high resolving power. Methodological improvements, as well as hyphenation of capillary IEF with other electrophoretic and chromatographic separation procedures, employed its versatility in studies of clinically important proteins, recombinant products, cell lysates, and other complex mixtures. The combination of capillary IEF with mass spectrometric detection is one of the major challenges for studying proteomics. Several reviews have appeared in this field[100–106] and a separate chapter summarizes the analytical advantages of the method.

IX. SPECIAL FEATURES IN THE PRACTICE AND THEORY OF IEF

Rosengren et al.,[107] described "a simple method in polyacrylamide gel slab for choosing optimum pH conditions for electrophoresis," which is a direct display of the titration curves of all the proteins present in a mixture. The first dimension consists of electrophoretically sorting the carrier ampholytes contained in the gel, thus resulting in a stationary pH gradient throughout the gel. The sample is analyzed by running it in a trough perpendicular to the first dimension and along the pH gradient. This generates electrophoretic pH titration curves that can be used to follow up genetic mutants of the proteins, macromolecule–ligand interactions, and macromolecule–macromolecule interactions. It is possible to determine directly the p*K* and pI values from titration curves (see a detailed description of the technique in the monograph of Righetti[42]).

The technique of chromatofocusing was first described by Sluyterman and co-workers.[108–110] They proposed that a pH gradient could be produced on an ion-exchange chromatography column by taking advantage of the buffering action of the charged group of the ion exchanger. If a buffer, initially adjusted to one pH value, is run through an ion-exchange column initially adjusted to a second pH value, a pH gradient is formed as if two buffers at different pH values were gradually mixed in the mixing chamber of a gradient maker. If such a pH gradient is used to elute proteins bound to the ion exchanger, the proteins elute in order of their isoelectric points. Furthermore, focusing effects take place, resulting in band sharpening, sample concentration and very high resolution. The technique is described in detail in Chapter 11.

The first theoretical description of the IEF process was made by theoretical modelling from Bier's group.[111,112] They generated theoretical pH gradients with mixtures of only two or three ampholytes of known electrochemical properties under a set of known physical parameters. The theoretical background of IEF was further studied, and model calculations were performed to understand the separation process with ampholytes in coated and uncoated capillaries, i.e., in the absence and presence of electroendosmosis.[113,114]

X. REVIEWS ON IEF

A review of carrier ampholytes has been published[115] and Rilbe has written an interesting autobiography.[116] In addition to proceedings of meetings, books covering general theoretical and methodological aspects of the technique[117] as well as its biomedical and biological applications[118] have been published. As a general reading, the book, *Electrokinetic Separation Methods*[119] was published in 1979 and, in 2001, *The Proteome Revisited: Theory and Practice of All Relevant Electrophoretic Steps*[120], which cover practically all aspects of electrophoresis in 21 chapters was published.

A host of reviews covering practically all aspects of IEF have also been published over the years. We would like to recommend some particularly interesting works about the historical moments behind the development of IEF and its sibling electromigration techniques. Of particular relevance are the manuals fully devoted to all facets of IEF, published in 1976 by Righetti and Drysdale[121] and about all of Righetti's fascinating writings on IEF that are difficult to surpass in both scientific value and literary language.[122] Rilbe has some equally fine retrospective writings[116,123] and Hjertén's excellent work about the development of electrophoresis in Uppsala[124] should be mentioned here. Recently several reviews have appeared by various authors.[100–106,125–127]

XI. CONCLUDING REMARKS

IEF has run through a long and intensive developmental process. As with any instrumental techniques, the progress of IEF possesses maxima and minima at times when either the technique was in favor above others or it suffered from difficulties in effective applications. The convenience of searching electronic databases for following the number of scientific papers published on IEF as a "concept" from the beginning to date allows us to view these "ups and downs." Certainly, the citation index does not provide the entire number of publications or even the number of cases when the technique was employed, however, it can be useful as a measure of interest in and importance of the technique (Figure 11). It is apparent that after a very short, very intensive increase in the late 1960s (when the technique itself was born) and after the renewed interest (the introduction of two-dimensional gel electrophoresis) towards the mid-1980s, a slow decrease can be seen from the 1990s. However, the overall annual production of publications is still high.

The technique of IEF is fully matured now and our work is completed at this point. Further we will await more innovations that will make IEF even more versatile and useful.

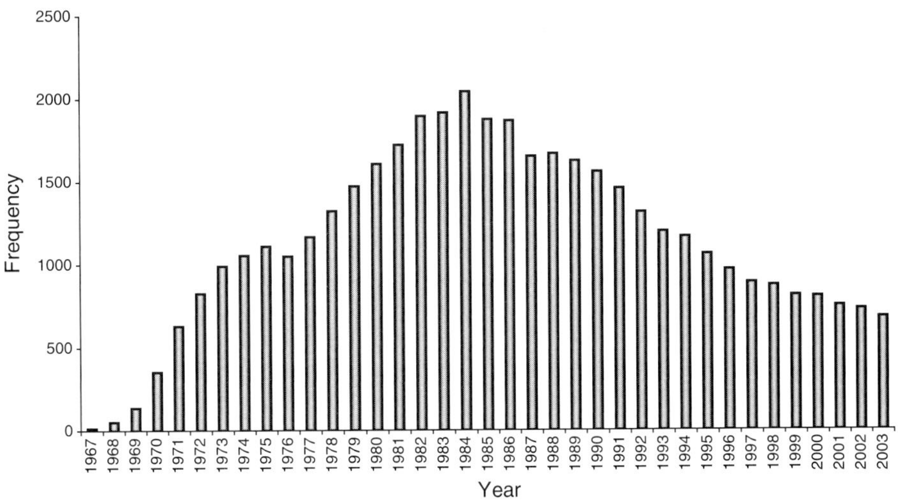

FIGURE 11 **The frequency of articles from 1967 dealing with IEF as a separation concept.**

REFERENCES

1. Kolin, A. Evolution of ideas in electrophoretic developments—selected highlights. In *Electrophoresis'82* (Stathakos, D., Ed.) de Gruyter, Berlin, pp. 3–48, 1983.
2. Volta, A. On the electricity excited by the mere contact of conducting substances of different kinds in a letter from Mr. Alexander Volta, F. R. S., Professor of Natural Philosophy at the University of Pavia. *Phil. Trans.* 90:403–431, 1800.
3. Reuss, F. F. Sur un nouvel effet de l'électricité galvanique. *Mém. Soc. Impériale Nat., Moscou* 2:327–337, 1809.
4. Helmholtz, H. Z. Studien über electrische Grenzschichten. *Ann. Phys. Chem.* 7:337–383, 1879.
5. Smoluchowski, M. Contribution to the theory of electro-osmosis and related phenomena. *Bull Int. Acad. Sci. Cracovie,* 3:184–199, 1903.
6. Gouy, G. Sur la constitution de la charge électrique à la surface d'un électrolyte. *J. Phys.* 9:457–468, 1910.
7. Chapman, D. L. A contribution to the theory of electrocapillarity. *Phil. Mag.* 6:475–481, 1913.
8. Stern, O. The theory of the electrolytic double-layer. *Z. Elektrochem. Angew. Phys. Chem.* 30:508–516, 1924.
9. Debye, P. and Hückel, E. The theory of electrolytes. I. Lowering of freezing point and related phenomena. *Physik. Z.* 24:185–206, 1923.
10. Arrhenius, S. A. Recherches sur la conductibilité galvanique des électrolytes (Investigations on the galvanic conductivity of electrolytes). Doctoral dissertation, Uppsala University, 1884.
11. Kohlrausch, F. Über Concentrations – Verschiebungen durch Electrolyse im Inneren von Lösungen und Lösungsgemischen. *Ann. Phys. Chem.* 62:209–239, 1897.
12. Lodge, O. On the migration of ions and an experimental determination of absolute ionic velocity. In Report of the 56th Meeting of the British Association for the Advancement of Science, pp. 389–413, 1886.
13. Tiselius, A. The moving boundary method of studying the electrophoresis of proteins. Doctoral dissertation, Uppsala University, 1930.
14. Picton, H. and Linder, S. E. Solution and pseudo-solution. Part I. *J. Chem. Soc.* 61:148–172, 1892.
15. Durrum, E. L. A Microelectrophoretic and microionophoretic technique. *J. Am. Chem. Soc.* 72:2943–2948, 1950.
16. Michl, H. *Hochspannungs Elektrophorese*, Thieme, Stuttgart, Germany, 1962.
17. Smithies, O. Zone electrophoresis in starch gels: group variation in the serum proteins of normal human adults. *Biochem. J.* 61:629–641, 1955.
18. Raymond, S. and Weintraub, L. S. Acrylamide gel as a separation medium for zone electrophoresis. *Science* 130:711, 1959.
19. Davis, B. J. and Orstein, L. A new high resolution electrophoresis method. A paper presented at The Society of the Study of Blood, N. Y. Acad. of Med., 1959.
20. Hjertén, S. Presented by A. Tiselius in Quarterly Report No. 1 to European Research Office (985/DU) US Department of the Army, Frankfurt/Main, Germany, APO 757, US Forces, 1960.
21. Hjertén, S. Agarose as an anticonvection agent in zone electrophoresis. *Biochim. Biophys. Acta* 53:514–517, 1961.
22. Ikeda, K. and Suzuki, S. Separating glutamic acid and other products of hydrolysis of albuminous substances from each other by electrolysis. US Patent No. 1, 015–891, 1912.
23. Williams, R. R. and Waterman, R. E. Electrodialysis as a means of characterizing ampholytes. *Proc. Soc. Exp. Biol. Med.* 27:56–59, 1929.
24. du Vigneaud, V., Irwing, G. W., Dyer, H. M. and Sealock, R. R. Electrophoresis of posterior pituitary gland preparations. *J. Biol. Chem.* 123:45–55, 1938.
25. Kolin, A. Separation and concentration of proteins in a pH field combined with an electric field. *J. Chem. Phys.* 22:1628–1629, 1954.

26. Kolin, A. Isoelectric spectra and mobility spectra: A new approach to electrophoretic separation. *P. Natl. Acad. Sci. USA* 41:101–110, 1955.
27. Kolin, A. Electrophoretic "line spectra". *J. Chem. Phys.* 23:407–408, 1955.
28. Kolin, A. In *Electrofocusing and Isotachophoresis: Proceedings of the International Symposium*, August 2–4, 1976, Hamburg, Germany (Radola, B. J. and Graesslin, D. Eds.) Walter de Gruyter, Berlin, pp. 3–33, 1977.
29. Friedli, W. and Schumacher, E. Über elektrophoretische Ionenfokussienung X. Die Analyse von Seltenen Erdgemischen. *Helv. Chim. Acta* 44:1829–1856, 1961.
30. Michl, H. Quantitative measurement of electrophoresis diagrams on filter paper. *Monatsh. Chem.* 83:210–220, 1952.
31. Righetti, P. G. Recent Developments in Titration Curves of Proteins by Isoelectric Focusing–Electrophoresis. In *Electrophoresis '81* (Allen, R. C. and Arnaud, P. Eds.) de Gruyter, Berlin, pp. 655–665, 1981.
32. Stahl, E. Gradient and low-temperature thin-layer chromatography. *Angew. Chem. Int. Ed.* 3:784–791, 1964.
33. Stahl, E. and Miller, J. pH-Gradient-Dünnschicht-Chromatographie von Benzodiazepinen. *J. Chromatogr.* 209:484–488, 1981.
34. Tate, M. E. Determination of ionization constants by paper electrophoresis. *Biochem. J.* 195:419–429, 1981.
35. Jokl, V., Dolejšová, J. and Matušová, M. Zone electrophoresis of organic acids and bases in water-alcohol solvents. *J. Chromatogr.* 172:239–248, 1979.
36. Tiselius, A. Stationary electrophoresis of ampholyte solutions. *Svensk kemisk tidskrift* 53:305–310, 1941.
37. Svensson, H. A discussion on the meaning of equivalent weights and transport (transference) numbers for amphoteric electrolytes, especially protolytes. *Sci. Tools* 3:30–35, 1956.
38. Svensson, H. Isoelectric fractionation, analysis and characterization of ampholytes in natural pH gradients. I. The differential equation of solute concentrations at a steady state and its solution for simple cases. *Acta Chem. Scand.* 15:325–341, 1961.
39. Svensson, H. Isoelectric fractionation, analysis and characterization of ampholytes in natural pH gradients. II. Buffering capacity and conductance of isoionic ampholytes. *Acta Chem. Scand.* 16:456–466, 1962.
40. Svensson, H. Isoelectric fractionation, analysis, and characterization of ampholytes in natural pH gradients. III. Description of apparatus for electrolysis in columns stabilized by density gradients and direct determination of isoelectric points. *Arch. Biochem.* (Suppl. 1):132–135, 1962.
41. Vesterberg, O. Synthesis and isoelectric fractionation of carrier ampholytes. *Acta Chem. Scand.* 23:2653–2666, 1969.
42. Righetti, P. G. *Isoelectric Focusing: Theory, Methodology and Applications.* Elsevier, Amsterdam, The Netherlands, pp. 15–31, 1983.
43. Vesterberg, O. Separation of proteins from carrier ampholytes. *Sci. Tools* 16:24–27, 1969.
44. Vesterberg, O. History of electrophoretic methods. *J. Chromatogr.* 480:3–19, 1989.
45. Vesterberg, O. and Svensson, H. Isoelectric fractionation, analysis and characterization of ampholytes in natural pH gradients. IV. Further studies on the resolving power in connection with the separation of myoglobins. *Acta. Chem. Scand.* 20:820–834. 1966.
46. Vesterberg, O. Isoelectric focusing of proteins. *Svensk kemisk tidskrift* 80:213–225, 1968.
47. Rilbe, H. and Pettersson, S. A Simple method for preparation of approximately constant density gradients in small columns. *Separ. Sci. Technol.* 3:535–549, 1968.
48. Jonsson, M., Pettersson, P. and Rilbe, H. Scanning isoelectric focusing in small density-gradient columns. I. Use of a standard spectrophotometer cuvette for focusing. chemical modification of proteins by migrating reactive ions. *Anal. Biochem.* 51:557–576, 1973.
49. Fredriksson, S. Scanning isoelectric focusing in small density-gradient columns 4. Use of deuterium-oxide for preparing density gradient and its effects on isoelectric points of proteins. *J. Chromatogr.* 108:153–167, 1975.

50. Rilbe, H. and Pettersson, P. Preparative isoelectric focusing in short density gradient columns with vertical cooling. In *Isoelectric Focusing,* (Arbuthnott, J. P. and Beeley, J. A. Eds.) Butterworths, London, pp. 44–57, 1975.
51. Vesterberg, O. The carrier ampholytes. In *Isoelectric Focusing* (Catsimpoolas, N., Ed.) Academic Press, London, pp. 53–76, 1976.
52. Pogacar, P. and Jarecki, R. Isoelectric focusing using inorganic acidic ampholytes In: *Electrophoresis and Isoelectric Focusing in Polyacrylamide gel,* (Allen, R. C. and Maurer, H. R. Eds.) de Gruyter, Berlin, pp. 153–158, 1974.
53. Grubhofer, N. and Borja, C. Synthesis of carrier ampholytes for isoelectric focusing containing sulfonic and phosphonic acid groups covering a wide pH range. In *Electrophoresis and Isoelectric Focusing: Proc. Int. Symp. 1976* (Radola, B. J. and Greassin, D. Eds.) de Gruyter, Berlin, pp. 111–120, 1977.
54. Williams, K. W. and Söderberg, L. A. Carrier ampholyte for isoelectric focusing. *Int. Lab.* 1:45–53, 1979.
55. Söderberg, L., Buckley, D., Hagström, G. and Bergström, J. The chemical properties of pharmalyte. In *Protides of the Biological Fluids,* (Peelers, H. Ed.) Pergamon Press, Oxford, pp. 687–691, 1979.
56. Vinogradov, S. N., Lowenkron, S., Andonian, H. R., Bagshow, J., Felgenhauer, K. and Pak, S. J. Synthetic ampholytes for the isoelectric focusing of proteins. *Biochem. Biophys. Res. Comm.* 54:501–506, 1973.
57. Charlionet, R., Martin, J. P., Sesboue, R., Madec, P. J. and Lefebvre, F. Synthesis of highly diversified carrier ampholytes—evaluation of the resolving power of isoelectric-focusing in the Pi system (alpha-1-antitrypsin genetic-polymorphism). *J. Chromatogr.* 176:89–101, 1979.
58. Charlionet, R., Morcamp, C., Sesboue, R. and Martin, J. P. Limiting factors for the resolving power of isoelectric-focusing in natural pH gradients. *J. Chromatogr.* 205:355–366, 1981.
59. Just, W. W. Synthesis of carrier ampholyte mixtures suitable for isoelectric fractionation analysis. *Anal. Biochem.* 102:134–144, 1980.
60. Righetti, P. G. and Hjertén, S. High-molecular-weight carrier ampholytes for isoelectric focusing of peptides *J. Biochem. Biophys. Methods* 5:259–272, 1981.
61. Luner, S. J. and Kolin, A. A new approach to isoelectric focusing and fractionation of proteins in a pH gradient. *Proc. Natl. Acad. Sci. USA* 66:898–903, 1970.
62. Lundahl, P. and Hjertén, S. Isoelectric focusing in free Ampholine™ solution and attempts at isoelectric focusing in pH gradients created in ordinary buffers. *Ann. New York Acad. Sci.* 200:94–111, 1973.
63. Troitzki, G. V., Savialov, V. P., Kirjukhin, I. F., Abramov, V. M. and Agitsky, G. J. Isoelectric focusing of proteins using a pH gradient by a concentration gradient of non-electrolytes in solution. *Biochim. Biophys. Acta* 400:24–31, 1975.
64. Vesterberg, O. Isoelectric focusing of proteins in polyacrylamide gels. *Biochim. Biophys. Acta* 257:11–30, 1972.
65. Vesterberg, O. Isoelectric focusing of proteins in thin layers of polyacrylamide gels. *Sci. Tools* 20:22–29, 1973.
66. Görg, A., Postel, W. and Westermeier, R. Ultra-thin-layer isoelectric-focusing in polyacrylamide gels on cellophane. *Anal. Biochem.* 89:60–70, 1978.
67. Hjertén, S. Free zone electrophoresis. *Chromatogr. Rev.* 9:122–219, 1967.
68. Hjertén, S. Zone electrophoresis, isoelectric focusing, and displacement electrophoresis (isotachophoresis) in carrier-free solution. In *Methods of Protein Separation,* Vol. 2, (Catsimpoolas, N. Ed.) Plenum, New York, pp. 219–231, 1976.
69. Bjellqvist, B., Ek, K., Righetti, P. G., Gianazza, E., Görg, A., Postel, W. and Westermeier, R. Isoelectric-focusing in immobilized pH gradients—Principle, methodology and some applications. *J. Biochem. Biophys. Methods* 6:317–339, 1982.
70. Gasparic, V., Bjellqvist, B. and Rosengren, Å. Manufacture of a pH-function for electrophoretic separation. Swedish Patent no. 7514049-1, 1975.

71. Rosengren, Å., Bjellqvist, B. and Gasparic, V. Manufacture of a pH-function for electrophoretic separation. US Patent no. 4,130,470, 1978.
72. Rosengren, Å., Bjellqvist, B. and Gasparic, V. Manufacture of a pH-function for electrophoretic separation. German Patent no. 2656162, 1981.
73. Altland, K. and Rossmann, U. Hybrid isoelectric focusing in rehydrated immobilized pH gradients with added carrier ampholytes: demonstration of human globins. *Electrophoresis* 6:314–325, 1985.
74. Righetti, P. G. *Immobilized pH Gradients: Theory and Methodology*. Elsevier, Amsterdam, 1990.
75. Righetti, P. G., Chiari, M., Casale, E. and Chiesa, C. Oxidation of alkaline immobiline buffers for isoelectric focusing in immobilized pH gradients. *Appl. Theor. Electrophoresis* 1:115–121, 1989.
76. Gaveby, B. M., Pettersson, E., Andrasko, J., Ineva-Flygare, L., Johannesson, U., Görg, A., Postel, W., Domscheit, A., Mauri, E. L., Pietta, E., Gianazza, E. and Righetti, P. G. Stable storage-conditions of immobiline chemicals for isoelectric-focusing. *J. Biochem. Biophys. Methods* 16:141–164, 1988.
77. Esteve-Romero, J., Simb-Alfonso, E., Bossi, A., Bresciani, F. and Righetti, P. G. Sample streaks and smears in immobilized pH gradient gels. *Electrophoresis* 17:704–708, 1996.
78. Smithies, O. and Poulik, M. D. Two-dimensional electrophoresis of serum proteins. *Nature* 177:1033, 1956.
79. Margolis, J. and Kenrick, K. G. Two-dimensional resolution of plasma proteins by combination of polyacrylamide disc and gradient gel electrophoresis. *Nature* 221:1056–1057, 1969.
80. Dale, G. and Latner, A. L. Isoelectric focusing of serum proteins in acrylamide gels followed by electrophoresis. *Clin. Chim. Acta* 24:61–68, 1969.
81. Macko, V. and Stegemann, H. Mapping of proteins by combined electrofocusing and electrophoresis. Identification of varieties. *Hoppe-Seyler's Z. Phys. Chem.* 350:917–919, 1969.
82. Stegemann, H. Proteinfraktionierungen in Polyacrylamid und die Anwendung auf die genetische Analyse bei Pflanzen. *Angew. Chem.* 82:640, 1970.
83. O'Farrell, P. High resolution two-dimensional electrophoresis of proteins *J. Biol. Chem.* 250:4007–4021, 1975.
84. Klose, J. Protein mapping by combined isoelectric focusing and electrophoresis of mouse tissues—Novel approach to testing for induced point mutations in mammals. *Humangenetik* 26:231–243, 1975.
85. Scheele, G. A. Two-dimensional gel analysis of soluble proteins. Charaterization of guinea pig exocrine pancreatic proteins. *J. Biol. Chem.* 250:5375–5385, 1975.
86. Anderson, N. L. and Anderson, N. G. High resolution two-dimensional electrophoresis of human plasma proteins *Proc. Natl. Acad. Sci. USA* 74:5421–5425, 1977.
87. Towbin, H., Staehelin, T. and Gordon, J. Electrophoretic transfer of proteins from polyacrylamide gels to nitrocellulose sheets: procedure and some applications. *Proc. Natl. Acad. Sci. USA* 76:4350–4354, 1977.
88. Aebersold, R., Leavitt, J., Saavedra, R. A., Hood, L-E. and Kent, S. B. H Electroblotting onto activated glass—high-efficiency preparation of proteins from analytical sodium dodecyl sulfate-polyacrylamide gels for direct sequence-analysis. *J. Biol. Chem.* 261:4229–4238, 1986.
89. Aebersold, R., Leavitt, J., Saavedra, R. A., Hood, L. E. and Kent, S. B. H Internal amino-acid sequence-analysis of proteins separated by one-dimensional or two-dimensional gel-electrophoresis after *in situ* protease digestion on nitrocellulose. *Proc. Natl. Acad. Sci. USA* 84:6970–6974, 1987.
90. Van der Sluis, P. J., Pool, C. W. and Sluiter, A. A. Press-blotting on gelatin-coated nitrocellulose membranes. A method for sensitive quantitative immunodetection of peptides after gel isoelectric focusing. *J. Immunol. Methods* 104:65–71, 1987.

3 THEORY AND SIMULATION OF ISOELECTRIC FOCUSING

T.L. SOUNART[a], P.A. SAFIER[b], AND J.C. BAYGENTS[b]

[a]*Sandia National Laboratories, Albuquerque, NM 87185-1411*
[b]*The University of Arizona, Tucson, AZ 85721*

I. PRINCIPLES OF ISOELECTRIC FOCUSING

Isoelectric focusing (IEF) is an electrophoretic separation scheme tailored to amphoteric compounds. IEF is used primarily to resolve mixtures of proteins and/or peptides. Similar to any other charged solute, an amphoteric compound translates under the action of an externally applied electric field—a process known alternatively as electromigration or electrophoresis. Owing to the chemical composition of amphoteric substances, their electrophoretic mobility is a function of pH: at low pH, the mobility is positive; at high pH, the mobility is negative. The generic relationship between electrophoretic mobility and pH is sketched for an amphoteric compound in Figure 1, where μ_E denotes the electrophoretic mobility, and the curve is drawn for a solution of (approximately) constant ionic strength. The isoelectric point (pI) is the pH at which the electrophoretic mobility of the compound is nil. At a given ionic strength, each amphoteric species evinces a different pI and, in the IEF scheme, these differences in pI serve as the basis to resolve separands.

Handbook of Isoelectric Focusing and Proteomics
D. Garfin and S. Ahuja, editors.

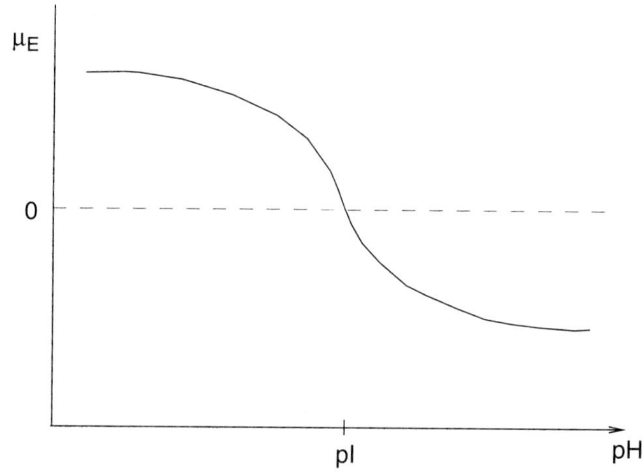

FIGURE 1 **Generic behavior of electrophoretic mobility versus pH for an amphoteric species at constant ionic strength.**

Since electrophoretic mobility is sensitive to pH, if an amphoteric species is placed in a buffer of non-uniform pH, the compound will move with an electrophoretic velocity that depends on position. The coupling between electrophoretic motion and pH leads to the focusing behavior sketched in Figure 2. Typically, IEF begins with separands distributed uniformly along the separation axis z. Under the circumstances shown in the figure, an amphoteric compound with a pI equal to the pH at z_0, migrates in the direction of the field if the substance is positioned at $z < z_0$. Conversely, the compound migrates against the field if located at $z > z_0$. At z_0, which is the position of the isoelectric point, the electrophoretic velocity of the compound vanishes. Thus a given amphoteric compound accumulates, or focuses, in the neighborhood of its isoelectric point.

The concept put forth in Figure 2 was first implemented by Kolin,[1] who focused proteins in the region between two counter-diffusing buffers of different pH values. Because pH gradients generated in this fashion are transient, the idea of IEF did not become broadly useful until Svensson[2–4] and Vesterberg[5,6] developed the requisite chemistries to generate stable pH gradients.

A. Steady Focusing and the Isoelectric Point

The purpose of this chapter is to discuss quantitative models of the IEF process. As we shall see later, constructing a thorough mathematical description of IEF is not a trivial matter. For the moment, however, we can restrict ourselves to the simple scenario depicted in Figure 2 and begin to uncover some of the underlying principles of IEF.

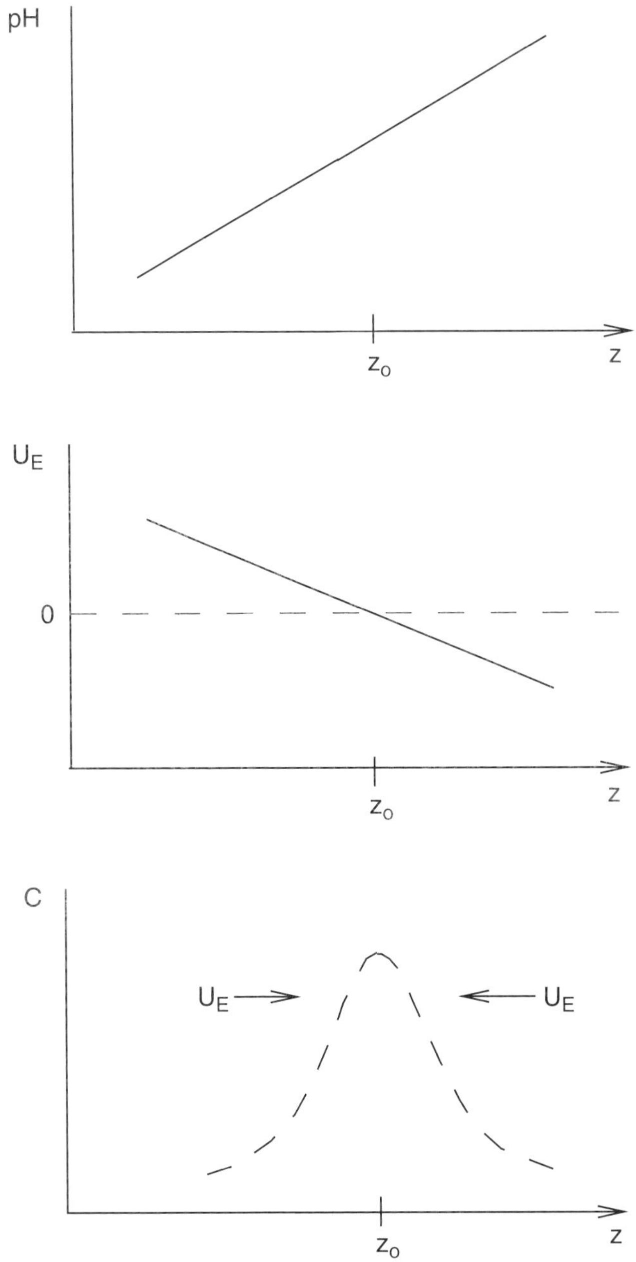

FIGURE 2 Isoelectric focusing of an amphoteric species in a local pH gradient. The electric field is taken to point from left to right (toward increasing z). Shown respectively in the panels, from top to bottom, are the pH and, for an amphoteric compound that focuses at position z_0, the electrophoretic velocity U_E and the concentration C as a function of z.

Before moving on to the topics of dynamics, it is worth noting that one of the reasons IEF is useful is that it evolves to a persistent steady state. For instance, the profile given by (6) arises irrespective of the initial condition on C—provided a stable pH gradient is generated. Figure 3 illustrates a hypothetical arrangement that yields no focusing of the amphoteric compound. Such a situation does not occur in practice; the electric field used to drive the electrophoretic transport of the separands to their respective pI values is also coupled with the generation of the pH profile. The electrode processes are such that the anodic end of the chamber is acidic and the cathodic end is basic. Therefore a steady arrangement of the electric field and the pH, as sketched in Figure 3, does not occur. Mosher et al.[8] discuss the generation of pH gradients in substantial detail and the reader should consult their monograph, as well as Chapter 4 of this text, to learn more.

B. Focusing Transients in a Steady pH Gradient

If we wish to examine the dynamics of the isoelectric focusing process, in general, we will have to consider the solution to a coupled set of non-linear balance laws that, at the very minimum, accounts for conservation of chemical species and satisfies Maxwell's equations for the electric field, which will be done in the later sections. In the interim it is instructive to consider the behavior of an amphoteric compound that is initially dispersed along a separation axis where the pH is fixed in space and time, as might be the case in a polyacrylamide gel with an immobilized pH gradient.[9] Under such a circumstance, Equation (1) suffices to describe the spatiotemporal evolution of the concentration of the ampholyte (amphoteric compound). Taking the pH and the electric field to be steady and decoupled from C, we can express the electrophoretic velocity of the amphoteric compound as

$$U_E = U_0 g(z) \tag{10}$$

where $g(z)$ is some known function of position and U_0 is indicative of the magnitude of the electrophoretic velocity. On substituting (10) into (2), we obtain the following dimensionless balance law from (1):

$$\frac{\partial C}{\partial t} + \frac{\partial}{\partial x}[g(x)C] = \frac{1}{Pe}\frac{\partial^2 C}{\partial x^2} \tag{11}$$

where x $(=z/L)$ is the axial position scaled on L and C and t are now dimensionless. The time and concentration scales are L/U_0 and M_{tot}/LS, respectively. $Pe \equiv U_0 L/D$ is a Peclet number weighing the relative importance of electrophoretic motion versus diffusion.

When the ampholyte balance is cast in the form of Equation (11), it is immediately evident that C will depend on the dimensionless parameter

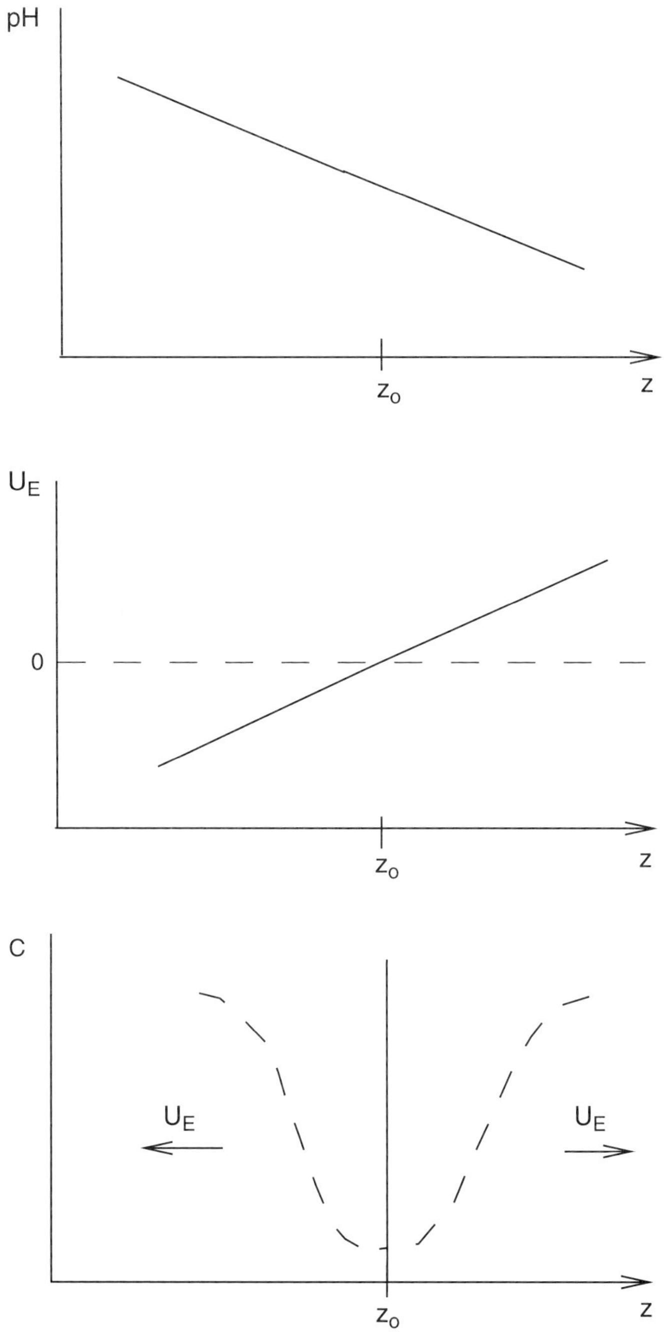

FIGURE 3 A local pH gradient that precludes focusing. The electric field is taken to point from left to right (toward increasing axial position z). Shown respectively in the panels, from top to bottom, are the pH and, for an amphoteric compound with a pI at z_0, the electrophoretic velocity U_E and the concentration C as a function of z.

Pe, i.e., $C = C(x, t; Pe)$. Typically Pe is quite large (>10^4) and, as will be seen, the Peclet number has a strong influence on the problem.

The function $g(x)$ accounts for spatial variations of the electrophoretic velocity of the ampholyte; $g(x)$ is a property of the mobility versus pH behavior of the particular ampholyte in question (e.g., Figure 1), as well as the electric field and the pH profile in the separation chamber. One must typically rely on numerical methods to construct solutions to (11), primarily because of the spatial dependence of $g(x)$. The form of $g(x)$ follows from the prescribed pH profile, the electric field, and the titration or mobility data for the ampholyte (e.g., Table 1). For the calculations presented in this section, the pH profile is taken to be linear in x, the electric field is uniform and the resultant $g(x)$ is closely approximated by cubic polynomials. Separate calculations will be shown for the proteins ferritin (Fer) and albumin (Alb). In the calculations for Fer, the pH is 2.0 at $x = 0$ and 10.0 at $x = 1$; for Alb, the pH varies from 3.0 to 11.15.

The numerical solutions to (11) at selected times for Fer and Alb are shown in Figures 4 and 5. The proteins are initially distributed uniformly over the interval $0 \leq x \leq 1$ and $Pe = 10^6$. Notice that by the time t moves past unity, the proteins have accumulated near their respective pI values, and at longer times the peaks are Gaussian. To obtain the time scales involved, consider the following. For the case of Fer with $Pe = 10^6$ (Figure 4), if the length of the separation axis L is 10cm, then the electric field is approximately 110V/cm and $t = 1$ corresponds to a dimensional time of about 5.3 min. For Alb (Figure 5) with $Pe = 10^6$ and $L = 10$ cm, the electric field is 40 V/cm and $t = 1$ corresponds to 2.9 min.

The results shown in Figures 4 and 5 were obtained with a numerical technique called flux-corrected transport[10] (FCT), which is designed to accommodate advectively dominated (i.e., high Peclet number) problems of the sort considered here. A comparison of results obtained by this method and another numerical technique (Petrov–Galerkin finite elements[12]) is shown in Figure 6. Notice that the FCT method captures the steep transitions more faithfully, which is an attribute that makes the method an attractive tool for the simulation of electrophoretic separations.[13]

The steady-state peaks for Alb at various Peclet numbers are shown in Figure 7. It is clear from these results that not only are the peaks Gaussian, but they narrow and their height increases with increasing Peclet number (Figure 8). More specifically, scaling arguments based on Equation (6) give that, for $Pe \gg 1$, the peak variance should change linearly with Pe^{-1} and C_{max} should vary as $Pe^{1/2}$. Figure 9 confirms this assertion, where plots of $Pe \times$ variance and $C_{max}/\sqrt{Pe}$ are markedly insensitive to changes in Pe.

Finally, it should be noted that, amongst the computational challenges posed by high Peclet numbers is the fine mesh required to resolve the narrow peaks that develop as the focusing proceeds. Since the variance scales as depends on $1/Pe$, the standard deviation of the peak (i.e., the peak width) scales as $1/\sqrt{Pe}$. At $Pe = 10^6$, if one were to discretize the

TABLE 1 Protein Properties[1]

	Net charge		
pH	**Hem**	**Alb**	**Fer**
2.0			18.8
2.5			13.2
3.0	68.5	58.0	8.5
3.3		44.5	
3.5	43.5	35.5	5.0
3.8		22.0	
4.0		13.0	2.0
4.1		8.0	
4.3		3.0	
4.5	25.5		0.0
4.8		0.0	
5.0			−2.0
5.4		−4.0	
5.5			−3.5
5.8		−6.1	
6.0	10.3		−5.5
6.5		−6.8	
6.8		−12.2	
7.0	0.0		−8.5
7.5			−10.0
7.8		−18.3	
8.0	−10.3		−11.5
8.5			−13.0
8.8		−24.4	
9.0	−20.5		−17.0
9.5			−19.5
10.0	−30.8	−32.0	−23.0
11.0	−50.0		
11.1		−48.0	
11.15		−64.0	
$e\omega_k$ (10^{-4} cm^2/V s)	0.265	0.231	0.126

[1]Hem and Alb data from Reference 8 and Fer data from Reference 22.

computational domain uniformly (equally spaced nodes), one could expect that no more than a few grid points would be located within the peak, even if as many as a thousand nodes were to be employed. To maintain computational efficiency, it is useful to employ a non-uniform grid, with many nodes positioned about the pI. Results obtained with uniform and non-uniform grids are shown in Figure 10, for Fer concentration at $Pe = 10^7$.

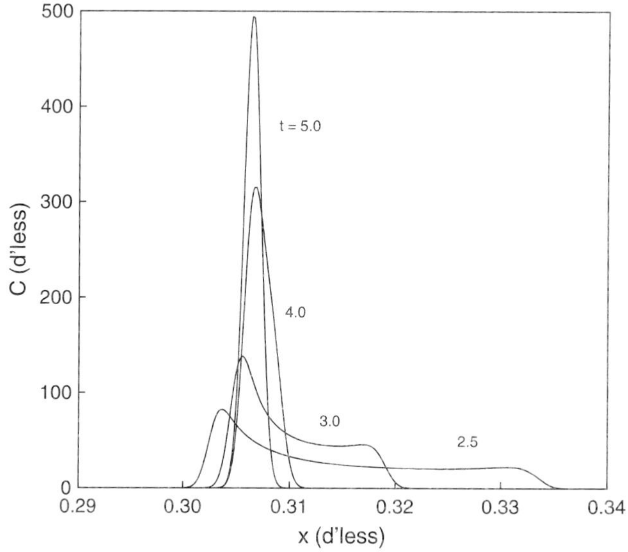

FIGURE 4 Ferritin concentration versus position for selected times. At $t = 0$, $C = 1$ for all x; $Pe = 10^6$. Numerical solutions obtained by the FCT method.[10] The spatial domain $x \in [0, 1]$ is discretized non-uniformly (1001 nodes); the computational mesh was generated with $\tau = 10$ in Equation (5-223) of Anderson.[11]

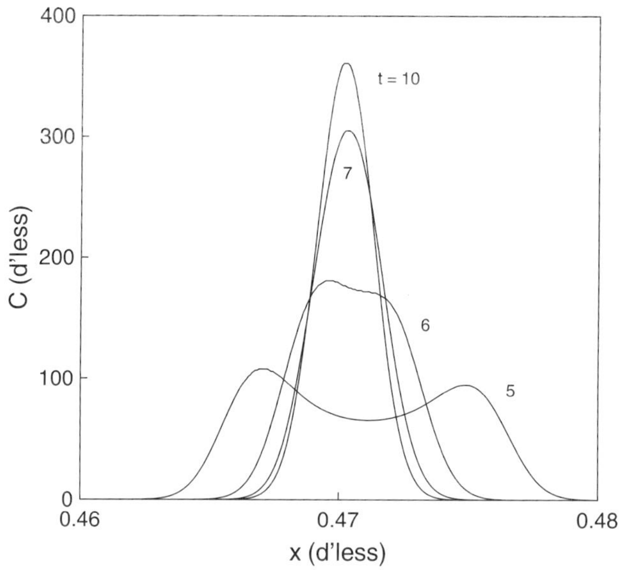

FIGURE 5 Albumin concentration versus position for selected times. Computational parameters are as in Figure 4.

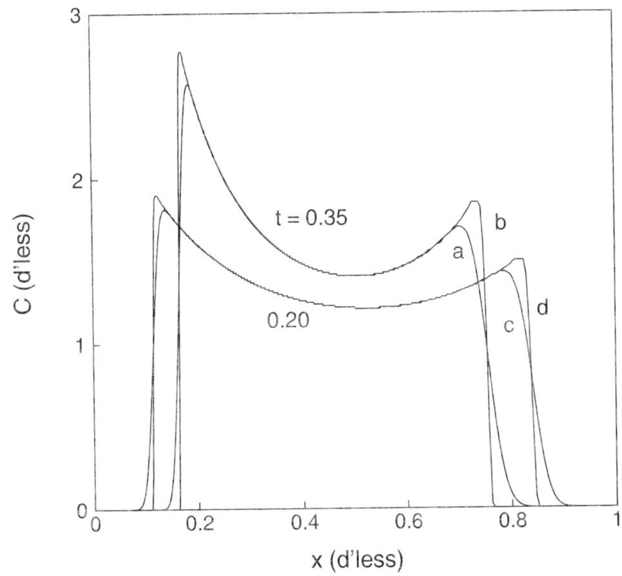

FIGURE 6 **Transient numerical solutions for concentration versus position: (a) and (c): Petrov-Galerkin finite elements[12]; (b) and (d): FCT.[10] At $t = 0$, $C = 1$ for all x. Concentration profiles are shown at dimensionless times of 0.2 and 0.35. The protein is albumin and $Pe = 10^6$. Each calculation uses 1001 nodes to discretize the spatial domain $x \in [0, 1]$. The non-uniform mesh was generated with $\tau = 10$ in Equation (5-223) of Anderson.[11]**

II. NUMERICAL SIMULATION OF IEF

The quantitative description we have presented so far is primitive in the sense that we have set aside the detailed physicochemical processes associated with IEF. For example, we have yet to account for the underlying development of a pH gradient suitable for IEF and we have ignored the important ionogenic mass-action equilibria that are characteristic of ampholytes. In this section we correct these deficiencies and lay out a comprehensive model of IEF. This is done chiefly by writing conservation relations for the electric charge and each of the chemical species that comprise the system (i.e., relationships akin to Equation (1)).

A. Balance Laws

A general set of balance laws governing the transport of ionic and neutral compounds in electrophoretic separations was developed in the 1980s,[14–16] and later detailed in a monograph by Mosher et al.[8] This coupled set of non-linear partial differential and algebraic equations includes an

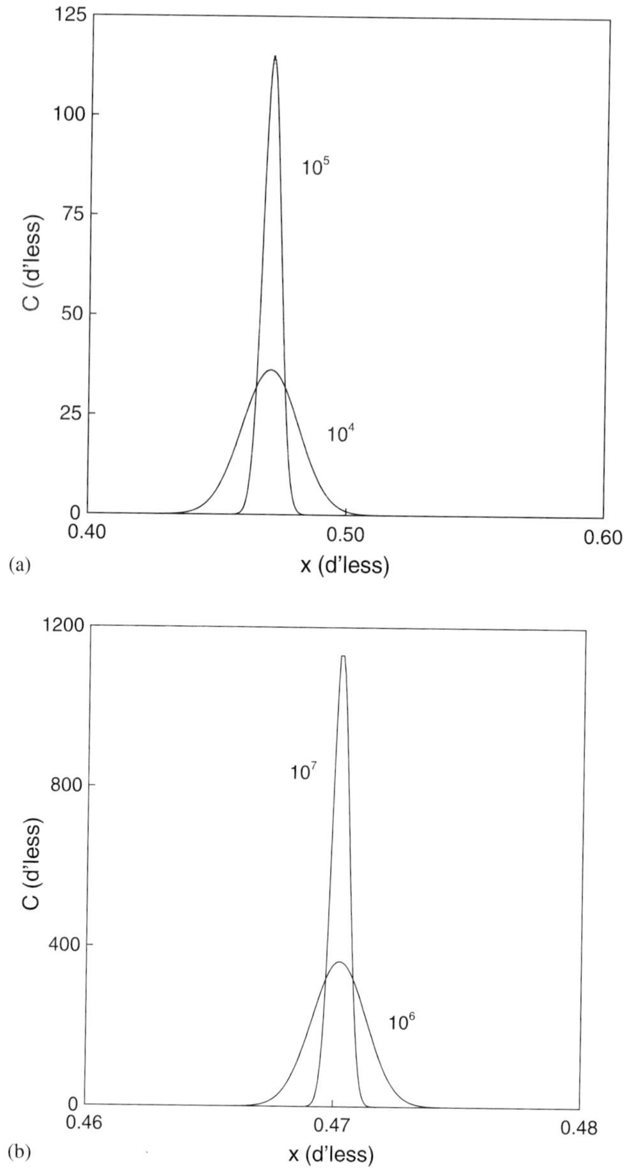

FIGURE 7 **Steady-state focusing of albumin at (a) $Pe = 10^4$ and 10^5 and (b) $Pe = 10^6$ and 10^7.**

unsteady electromigration–diffusion equation for each solutal component, a charge balance, the electroneutrality approximation, expressions for ionogenic dissociation–association equilibria, and a model for calculating protein mobilities as a function of pH and ionic strength. The balance laws

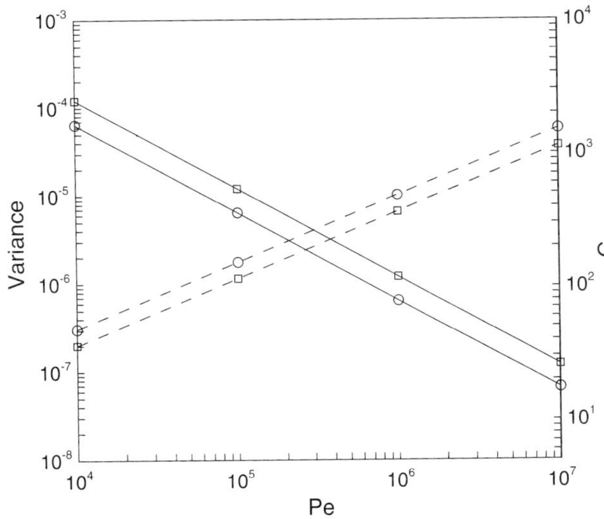

FIGURE 8 **Scaled peak variance (—) and scaled C_{max} (---) versus *Pe*: (○) Fer; (□) Alb. Remaining computational parameters are as specified in Figure 4.**

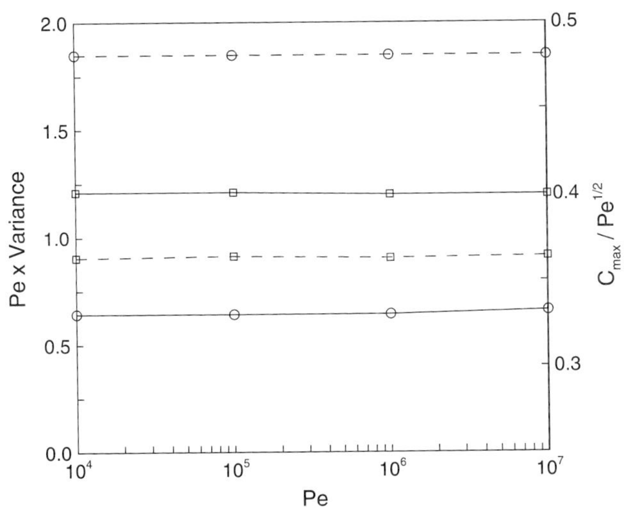

FIGURE 9 ***Pe* × variance (—) and $C_{max}/\sqrt{Pe}$ (---) versus *Pe*: (○) Fer; (□) Alb. Remaining computational parameters are as specified in Figure 4.**

are summarized here using the notation of related works on the dynamics of electrophoretic separations.[13,17]

Because dissociation–association reactions are fast compared with the mass transport, ion concentrations are constrained by a coupled set

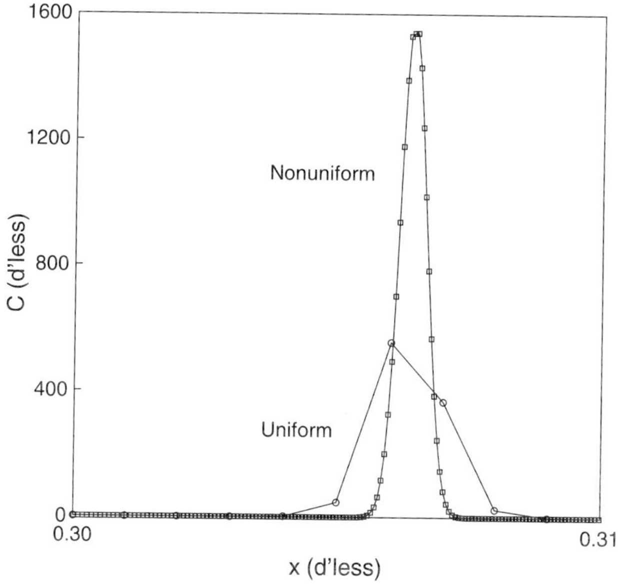

FIGURE 10 **Steady-state concentration versus position obtained from non-uniform and uniform computational grids. The protein is ferritin and $Pe = 10^7$. Each calculation uses 1001 nodes to discretize the spatial domain $x \in [0, 1]$. The non-uniform mesh was generated with $\tau = 10$ in Equation (5-223) of Anderson.[11]**

of mass-action relations. These include the dissociation of water, viz.

$$K_W \equiv [H^+][OH^-] = 10^{-14} M^2, \tag{12}$$

and the ion dissociation–association equilibria for M solutal components. If the neutral form of the kth component A_k^0 is protonated or deprotonated to form P_k cations and N_k anions, then the mass-action relations for ions of valence z are

$$A_k^z \rightleftharpoons H^+ + A_k^{z-1}, \qquad \begin{cases} z = -N_k+1, \ -N_k+2, \ldots, +P_k \\ k = 1, 2, \ldots, M \end{cases}$$

which are characterized by the equilibrium constants

$$K_k^z \equiv \frac{[H^+] n_k^{z-1}}{n_k^z}, \qquad \begin{cases} z = -N_k+1, \ -N_k+2, \ldots, +P_k \\ k = 1, 2, \ldots, M \end{cases} \tag{13}$$

where n_k^z is the concentration of subspecies A_k^z. Local neutrality prevails on length scales large compared with κ^{-1}, the Debye screening length, and thus for IEF,

$$0 = \sum_{k=1}^{M} \bar{z}_k C_k + [\mathrm{H}^+] - \frac{K_\mathrm{W}}{[\mathrm{H}^+]} \tag{14}$$

where C_k is the concentration and $\bar{z}_k$ is the effective valence of the kth component, i.e.,

$$C_k \equiv \sum_{z=-N_k}^{+P_k} n_k^z, \quad k=1, 2, \ldots, M \tag{15}$$

and

$$\bar{z}_k \equiv \frac{\sum_{z=-N_k}^{+P_k} z n_k^z}{C_k}, \quad k=1, 2, \ldots, M \tag{16}$$

To account for local variations of the solutal concentrations, a mass balance is written for each component, viz.

$$\frac{\partial C_k}{\partial t} = -\nabla \cdot \mathbf{f}_k = -\nabla \cdot [(\bar{\mu}_k^{\mathrm{e}} \mathbf{E} + \mathbf{v}) C_k - D_k \nabla C_k], \quad k = 1, 2, \ldots, M \tag{17}$$

where t is the time, $\mathbf{v}$ the fluid velocity, $\mathbf{E} = -\nabla\phi$ the local electric field, and $D_k = \omega_k k_\mathrm{B} T$ the diffusivity, with $k_\mathrm{B}T$ being the Boltzmann temperature and ϕ the electric potential; $\mathbf{f}_k$, ω_k, and $\bar{\mu}_k^{\mathrm{e}}$ are, respectively, the flux, the hydrodynamic mobility (taken here to be independent of component sub-speciation), and the effective electrophoretic mobility of component k. If Mp denotes the number of protein species,

$$\bar{\mu}_k^{\mathrm{e}} = \begin{cases} [f(\kappa a_k)/(1+\kappa a_k)] e\bar{z}_k \omega_k, & k = 1, 2, \ldots, M_\mathrm{p} \\ e\bar{z}_k \omega_k, & k = M_\mathrm{p}+1, M_\mathrm{p}+2, \ldots, M \end{cases} \tag{18}$$

where e is the charge on a proton (1.6×10^{-19} C), and a_k the (Stokes) radius of protein k; $f(\kappa a_k)$ is Henry's f function.[18]

The motion of the aqueous electrolyte is governed by the Navier–Stokes equations for incompressible flow, i.e.,

$$\rho \left[\frac{\partial \mathbf{v}}{\partial t} + \mathbf{v} \cdot \nabla \mathbf{v} \right] = -\nabla p + \mu \nabla^2 \mathbf{v} + \varepsilon_0 \varepsilon \nabla \phi \cdot \nabla \nabla \phi \tag{19}$$

$$\nabla \cdot \mathbf{v} = 0 \tag{20}$$

where ρ, μ, and ε are the fluid density, viscosity, and relative permittivity, respectively; p is the pressure and ε_0 is the permittivity of free space. The last term on the RHS of Equation (19) accounts for the electrical (Maxwell) stresses.

Since charge must be conserved, the governing equations are closed by combining the ion balances to obtain an equation for ϕ, viz.

$$\nabla \cdot (\sigma \nabla \phi) = -\nabla \cdot \mathbf{i}_D$$
$$= -ek_B T \nabla \cdot \left(\sum_{k=1}^{M} \omega_k \nabla \bar{z}_k C_k + \omega_H \nabla [H^+] - \omega_{OH} K_W \nabla [H^+]^{-1} \right) \quad (21)$$

where

$$\sigma = e^2 \left[\sum_{k=1}^{Mp} \frac{f(\kappa a_k)}{1+\kappa a_k} \overline{z_k^2} \omega_k C_k + \sum_{k=Mp+1}^{M} \overline{z_k^2} \omega_k C_k + \omega_H [H^+] + \omega_{OH} \frac{K_W}{[H^+]} \right] \quad (22)$$

is the local electrical conductivity and $\mathbf{i}_D$ is the diffusion current density, implicitly defined in Equation (21); $\overline{z_k^2}$ is the mean square valence of the kth component, i.e.,

$$\overline{z_k^2} \equiv \frac{\sum_{z=-N_k}^{+P_k} z^2 n_k^z}{C_k}, \qquad k = 1, 2, \ldots, M \quad (23)$$

Note that for components that may undergo many protonation or deprotonation reactions (e.g., proteins), dissociation–association equilibrium constants are not necessarily available. In such cases, subspecies concentrations are not calculated and, in place of Equations (16) and (23), effective and mean square valences are determined from titration data[8] (e.g., Table 1).

B. Initial and Boundary Conditions

Electroosmosis is often suppressed in IEF separations to eliminate dispersion from fluid motion in the inherently non-uniform electrolytes.[19] Consequently, all boundary conditions on the fluid velocity approach zero, and **v** vanishes in the separation channel. Initial conditions for the ionic concentrations are approximately cross-sectionally uniform, and so for electrically insulating channel walls, **E** has no transverse component. Cross-sectional uniformity is thus maintained throughout the separation, so typically the problem formulation need only be solved in one spatial (axial) dimension. If x is the axial coordinate, then for a constant voltage separation, the boundary conditions on $\phi(x, t)$ are

$$\phi(0, t) = V \quad (24)$$

$$\phi(L, t) = 0 \quad (25)$$

where V is the applied potential and L is the separation channel length. Initial and boundary conditions on the component concentrations vary with IEF technique, and will be discussed for each simulation presented in section III.

C. Numerical Implementation

The general electrophoresis model described above is solved here numerically for several IEF separation conditions applied in practice. Equation (17) is solved in the axial spatial dimension using the fifth-order-accurate Runge-Kutta-Fehlberg (RKF) algorithm for the time step and a second-order central spatial discretization. The solution of this equation is coupled to the remainder of the balance laws, which must be solved at each time step. In one dimension, Equation (21) reduces to an explicit expression for the electric field, viz.

$$E = \frac{i - i_{\mathrm{D}}}{\sigma} \tag{26}$$

where E and i_{D} are the axial components of $\mathbf{E}$ and $\mathbf{i}_{\mathrm{D}}$, respectively, and i is the current density given by

$$i = \frac{V}{\int_0^L 1/\sigma(x, t)\partial x} \tag{27}$$

The derivatives in i_{D} are calculated with second-order central finite differences, and the integral in Equation (27) is numerically integrated with a simple trapezoidal rule. These and the algebraic equations (12)–(16) are solved at each step in the RKF algorithm for Equation (17). This numerical scheme has been discussed at length elsewhere.[8,16]

III. ILLUSTRATIVE SIMULATIONS OF IEF

Commercial IEF buffers consist of dozens of amphoteric compounds with isoelectric points (pI) spanning the pI range of the analytes to be focused. If an initially uniform IEF buffer is confined in an electric field, a pH gradient is established as the ampholytes migrate to their pI. The buffer is confined in practice either by a closed column or more commonly by bounding the ampholytes between a strong acid and a strong base.

To illustrate IEF buffer concentration evolution during pH gradient formation, IEF is first simulated with a simple eight-component buffer. Each component has two pK values with $\Delta pK = 2$, mobility $e\omega_k = 3.0 \times 10^{-4}\,\mathrm{cm^2/V\,s}$, and pI values ranging from 3 to 10 in increments of 1. $V = 500\,\mathrm{V}$, $L = 5\,\mathrm{cm}$, all concentrations are initially uniform at 2.78 mM, and the column ends are impermeable to the buffer. These conditions approximate a closed channel filled with the ampholytes the ends or, an open channel with each end submerged in electrode reservoirs of strong acid and base. The time evolution of the pH and buffer distribution are shown in Figure 11. At the initial pH of 6.5, half the ampholytes are positively charged and half are negative, and so half initially migrate toward the anode and half toward the cathode (panel a). The positive

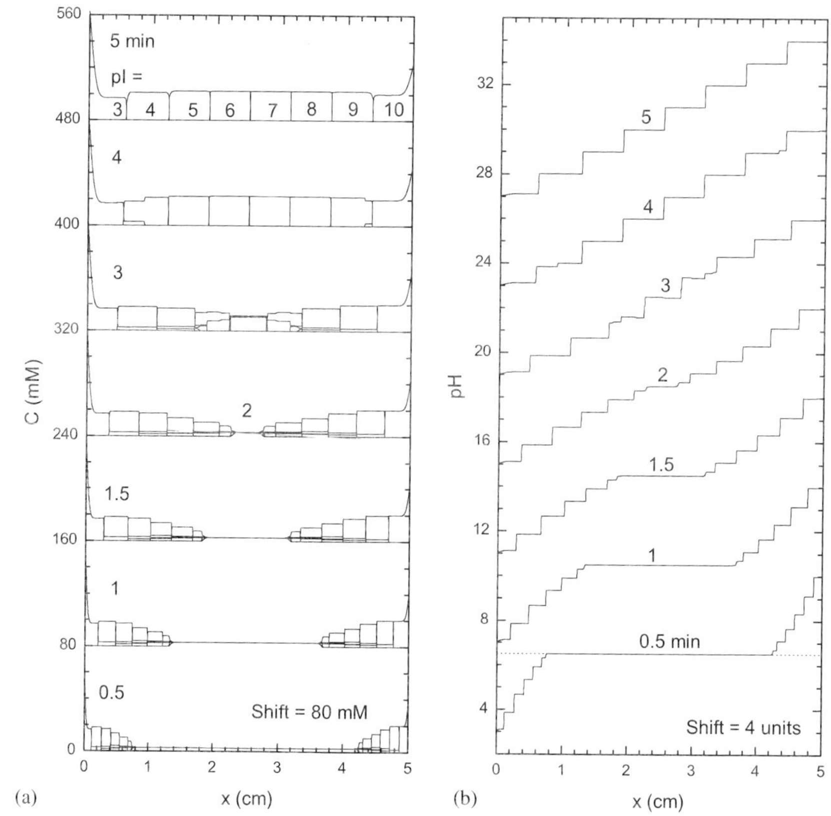

FIGURE 11 Dynamics of IEF buffer composed of eight carrier ampholytes with pI = 3–10, $\Delta pK = 2$, and $e\omega_k = 3.0 \times 10^{-4}\,cm^2/Vs$: (a) ampholyte concentrations; (b) pH. $V = 500$ V; $L = 5$ cm. All concentrations are initially uniform at 2.78 mM, and column ends are impermeable to the buffer. Dotted line represents the initial field. Plots are shifted as noted for clarity. Anode is to the left.

species leave the region near the anode as the negative species accumulate, and the charge is balanced by an increase in hydrogen ions. This shifts the equilibrium of the negative species towards their neutral state, reducing the effective mobility. Electromigration and thus accumulation stops when the pI is reached, so the most acidic component (pI = 3) builds up on the anode until the pH is 3. The component of pI = 4 is then forced to accumulate in a zone adjacent to this component because the charge of the pI = 4 component changes sign in the region of pH = 3, and so on for each ampholyte in the pI series. Therefore as the more acidic/basic species migrate toward the anode/cathode and accumulate in successive zones, the pH decreases/increases (panel b) in steps as the pI is approached for each component, at which point electromigration ceases and a steady-state distribution is established. The nonlinear coupling between the transport of each component and the electric field and pH leads to a complicated approach to steady state involving two zones for each component converging into one. Because only eight components comprise this IEF buffer, the pH gradient is not continuous. The channel is divided into eight stationary zones of equal length, with each zone composed of a uniform plug of each neutral ampholyte.

In the commercial IEF buffers used in practice, a continuous and approximately linear pH gradient forms because a sufficient number of ampholytes are included to create zones of length scale of the order of the characteristic diffusion length; the continuum results from many overlapping Gaussian-shaped ampholyte zones. Mao et al.[20] demonstrated that the behavior of Pharmalyte 3–10 can be predicted reasonably well with a buffer composed of 140 ampholytes of $\Delta pK = 2$, $e\omega_k = 3.0 \times 10^{-4}\,\mathrm{cm^2/Vs}$, and pI values ranging from 3 to 9.95 in increments of 0.05. Here we simulate the dynamics of 2% Pharmalyte 3–10 in the same electric field and channel as described above, with 141 of the same ampholytes (pI values from 3–10). All ampholyte concentrations are initially uniform at 0.16 mM. Results are presented in Figure 12. The pH evolves as for the eight component buffer, but the distribution is continuous (panel a). The dual-peak approach to steady-state ampholyte distributions is clearly seen in the focusing of ampholyte 71 (panel b). Each peak has the same height because this is the central ampholyte, and transport in the column is approximately symmetric. The ampholytes closer to the column ends are focused essentially at one peak during most of the development, with only very small secondary peaks forming transiently. Note that although a steady-state buffer distribution has not quite been realized in 5 min, the linear pH distribution has been formed. The conductivity is diminished as the pI of each carrier ampholyte is approached and charge carriers are depleted (panel c). This occurs initially at the channel ends where accumulation/depletion of ampholytes begins, and propogates towards the center of the column with time. The electric field scales inversely on σ (panel d, Equation (26)), increasing where charge is depleted to satisfy electroneutrality. The magnitude of the proportionality between E and $1/\sigma$ diminishes however, as the current density is reduced by a factor of 4 during IEF buffer focusing (Figure 16). This occurs because of the increase in channel resistance as the ampholytes are neutralized at their pI. The current thus provides a sensitive metric for pH gradient development, and shows an approximate steady state after about 5 min—the time required to produce a linear pH gradient.

The dynamics of the electric field distribution (Figure 12d) is key to understanding the dual-peak migration toward focused ampholyte zones. As charge is initially depleted at the channel ends, the electric field increases there and decreases in the center. The ampholytes thus migrate toward their pI from high-field regions at the ends to a low-field region in the center, leading to immediate accumulation into peaks at the channel

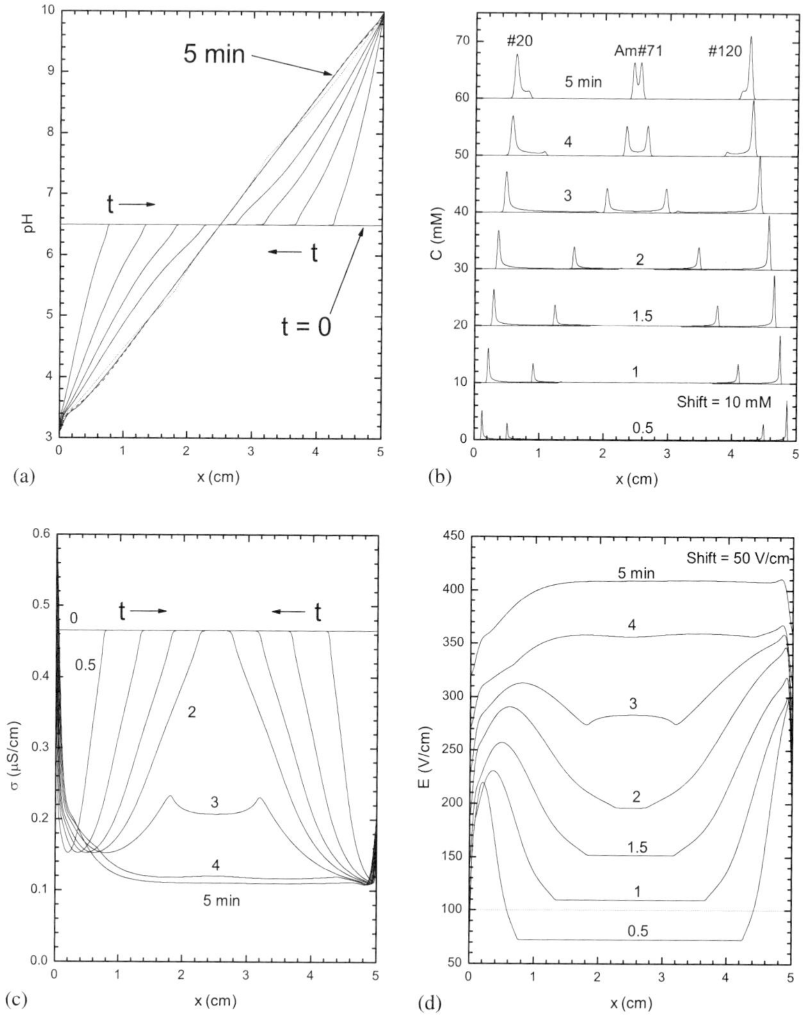

FIGURE 12 Simulation of 2% Pharmalyte 3–10 carrier buffer focusing: (a) pH; (b) selected carrier ampholytes 20, 71, 120; (c) conductivity; (d) electric field (magnitude) at selected times 0, 0.5, 1, 1.5, 2, 3, 4, 5 min. Carrier buffer is composed of 141 ampholytes with $\Delta pK = 2$ and $e\omega_k = 3.0 \times 10^{-4}$ cm^2/V s and pI = 3–10. All ampholyte concentrations are initially uniform at 0.16 mM. All other conditions as in Figure 11.

ends regardless of their pI values. The peak heights begin to diminish with time because as the ampholytes migrate, the electric field disturbances propagate toward the center of the channel, and the field gradient diminishes. As the ampholytes get closer to their pI, their transport becomes dominated more by pH variations than electric field variations, and they are ultimately immobilized in a single concentrated zone at the isoelectric point.

If dilute analytes are included in an IEF buffer, they are separated as they focus at their pI in the developing pH gradient. Because the analytes are dilute relative to the carrier buffer, they are essentially uncoupled from the electric field and other ionic transport. Figure 13 shows the focusing of five amphoteric dyes in the Pharmalyte buffer and channel just described. The initial uniform concentration of each dye is 1 μM, and the relevant properties of each dye are provided in Table 2. In Figure 13a, the time evolution of two dyes (D1 and D2) of the same electrophoretic mobility are shown. D2 focuses faster and sharper than D1 because it has a lower Δp*K*, and as such it is fully dissociated over a larger pH range. The effective mobility of this analyte increases sharply away from the pI, leading to efficient focusing and high resolution. Figure 13b shows the dynamics of two dyes (D3 and D4) with Δp*K* = 5.5. The D4 zone is spread over almost a full pH unit and requires 80 min to focus because of

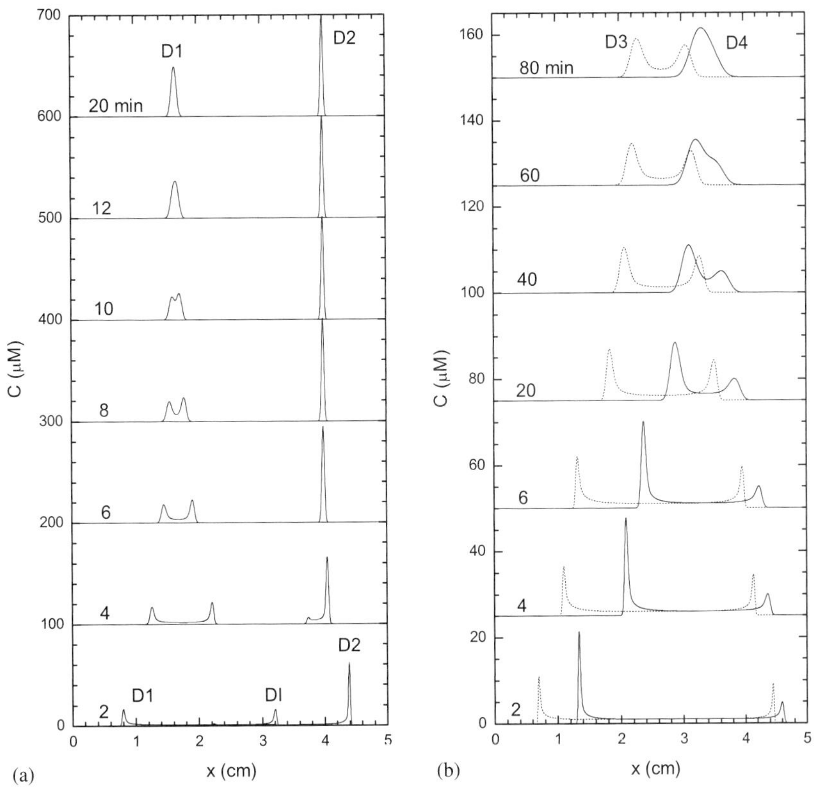

FIGURE 13 **Focusing of four amphoteric dyes in 2% Pharmalyte 3–10: (a) two dyes (D1, D2) with Δp*K* ≤3, eω_k = 3.0×10^{-4} cm^2/V s; (b) two dyes (D3, D4) with Δp*K* = 5, eω_K = 3.0 (—) and 1.0 (·····) × 10^{-4} cm^2/V s. All dye concentrations are initially uniform at 1 μM. All other conditions as in Figure 12.**

TABLE 2 Physicochemical parameters for non-linear simulations

Component	pK_1	pK_2	$e\omega_k$ (10^{-4} cm^2/V s)	Reference
pI 5.30 dye (D1)	3.70	6.90	3.0	20
pI 8.60 dye (D2)	7.68	9.52	3.0	20
pI 6.75 dye (D3)	4.02	9.49	1.0	21
pI 6.75 dye (D4)	5.01	10.50	3.0	21
H_3PO_4	2.00		3.67	20
Na^+			5.19	8
H^+			36.27	8
OH^-			19.87	8

the high Δpk values. D3 has not been separated even after 80 min because the electrophoretic mobility is a factor of 3 lower than that of D4. The efficacy of IEF is reduced as the analyte ΔpK increases and the electrophoretic mobility decreases. Note that during the dual-peak approach to steady state, the peaks are initially sharper and more concentrated than in the focused peak. This results from the transient initial increase in the electric field (three times the average) near the channel ends (Figure 12b) as described previously.

In Figure 14, IEF separation of three proteins is demonstrated. The initial uniform concentration of hemoglobin (H), albumin (A), and ferritin (F) is 1 μM, and input parameters are listed in Table 1. All proteins are separated, including albumin and ferritin, which have pIs differing only by 0.3 units. Hemoglobin is focused first (8 min), followed by albumin (20 min), and finally ferritin (40 min). The breadth of the zones and final concentration follows the same ordering, with hemoglobin concentrated by a factor of 400—the most of all analytes (including the dyes). The ordering of the focusing efficacy follows the ordering of the mobilities of the proteins (Table 1), and may also be influenced by the distribution of pK values.

In practice, an IEF focusing space is often created by bounding the carrier ampholytes and analytes in a channel between regions of strong acid and base.[20] This technique is simpler to implement than a closed channel, and has been employed in capillary IEF (CIEF) to transport a small mobile focusing space along a capillary to a detector, in a protocol similar to capillary zone electrophoresis (CZE) or isotachophoresis (ITP).[23] To demonstrate this technique, an IEF separation is simulated in 2% Pharmalyte bounded by H_3PO_4 at the anode and NaOH at the cathode in a 2 cm open channel at 200 V. The initial condition is shown in Figure 15a. H_3PO_4 (100 mM) fills the first 0.5 cm of the anodic end of the channel, and 40 mM NaOH fills 0.5 cm of the cathodic end, leaving a focusing space of length $L_f = 1$ cm in the center of the channel. The 141

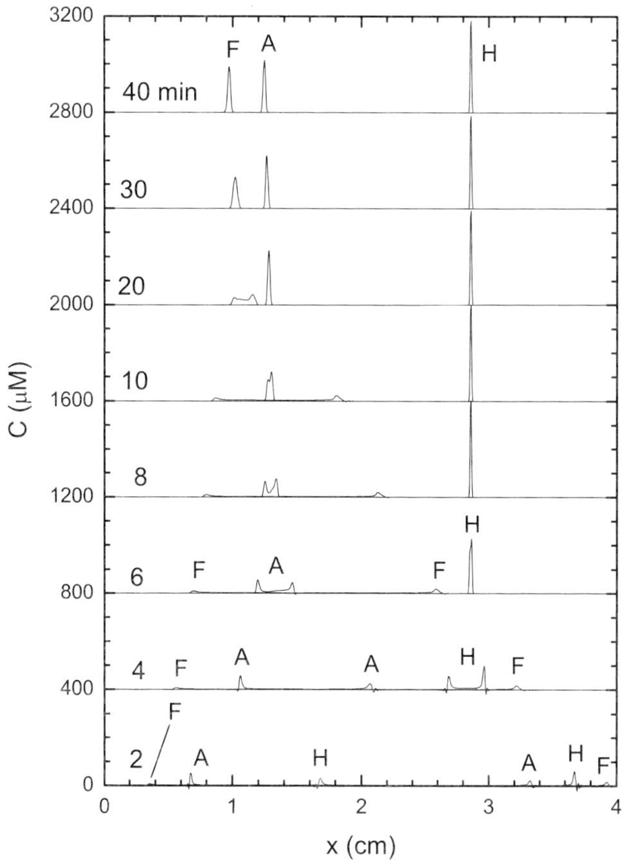

FIGURE 14 **Focusing of three proteins (hemoglobin H, albumin A, ferritin F) in 2% Pharmalyte 3–10. All protein concentrations are initially uniform at 1 μM. All other conditions as in Figure 12.**

ampholytes representing Pharmalyte 3–10 (0.16 mM), the dyes D1 and D2 (1 μM), and the three proteins (1 μM) initially fill the center 1 cm of the channel. Figure 15a shows the transient behavior of the carrier electrolytes. The strong electrolyte zones bounding the IEF buffer create impermeable barriers for the ampholytes because the electrophoretic velocity of even those with the lowest and highest pI reverses direction (from sign change) upon entering the low and high pH regions, respectively. The inset in Figure 15a, shows the ampholyte with pI = 3.0 (lowest) building up at the position of the strong-acid zone boundary. The focusing space behavior is essentially the same as for impermeable column ends (cf. Figure 12), but the strong acid and (particularly) base zones migrate slightly, causing the focusing space to drift during separation.[24]

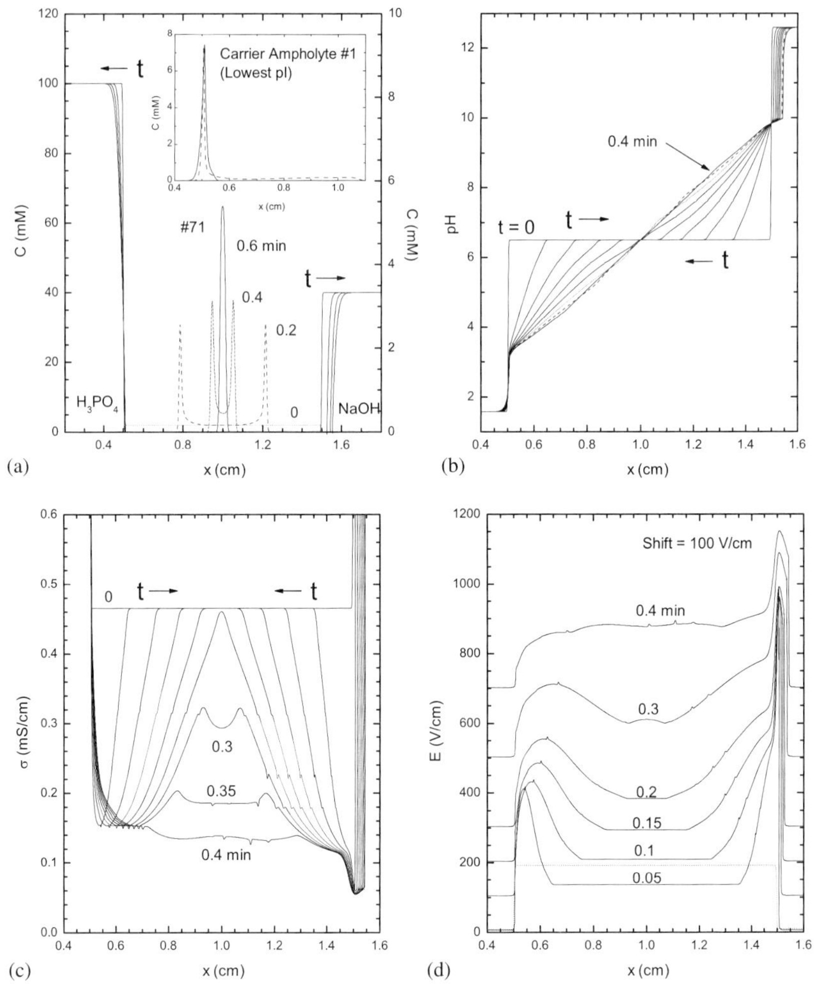

FIGURE 15 **2% Pharmalyte 3–10 focusing space bounded by strong acid H_3PO_4 and base NaOH. (a) Buffer zones at selected times 0, 0.2, 0.4, and 0.6 min. Concentration of H_3PO_4 and NaOH on left axis, ampholyte 71 on right axis; inset shows ampholyte 1 at 0.2 (---) and 0.6 (—) min. (b) pH; (c) conductivity; (d) electric field (magnitude) at selected times in the range 0–0.4 min in increments of 0.05 min. $L = 2$ cm; $V = 200$ V; column ends are open to H_3PO_4 and NaOH reservoirs. All carrier ampholyte concentrations are initially uniform at 0.16 mM between $x = 0.5$ and 1.5 cm, and zero elsewhere. All other conditions as in Figure 12.**

The migration of the NaOH zone also creates a dip in the conductivity (panel c) and a peak in the electric field (panel d) near that boundary. After 3s, the electric field near the NaOH zone boundary is over five times the field in the central plateau region where the buffer concentrations are still uniform.

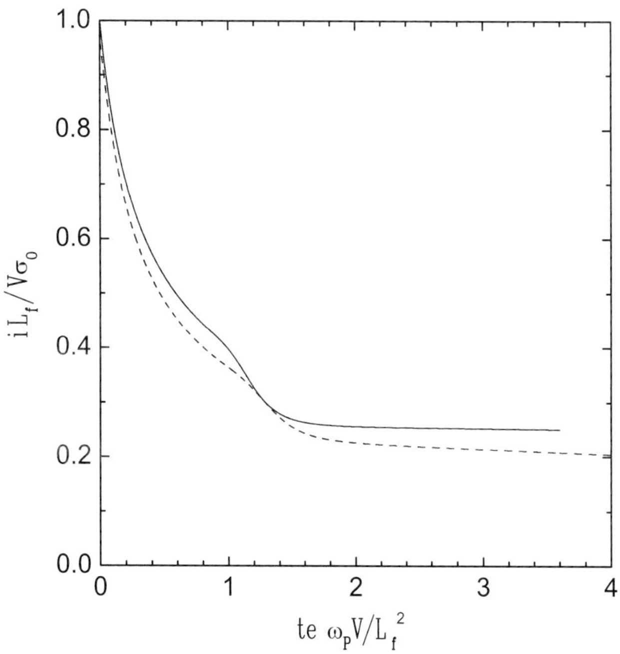

FIGURE 16 **IEF current density kinetics for 2% Pharmalyte 3–10 in closed channel (—) and bounded by strong acid and base in open channel (---). All conditions for closed channel are as in Figure 12; open channel as in Figure 15. σ_0 is the initial conductivity of 2% Pharmalyte 3–10, and $e\omega_k = 3.0 \times 10^{-4}$ cm²/V s is the mobility of the Pharmalyte ampholytes.**

Comparison of Figure 12 and 15 shows that in this simulation, the pH gradient develops an order of magnitude faster than in the closed column simulation. The faster separation is obtained because the channel is shorter and the average field in the focusing space is higher; the focusing time scales on $L_f^2/(e\omega_k V)$, which is a factor of ten higher in the closed channel simulation. Note that the scaled current density decay is similar to that in the closed column when plotted against non-dimensional time (focusing time scale) (Figure 16).

Figure 17 shows the IEF separation of five analytes in the 2% Pharmalyte buffer bounded by H_3PO_4 and NaOH. Analyte focusing behavior is similar to that in the closed channel (cf. Figures 13 and 14), with essential differences resulting only from the shorter focusing space and higher electric field. The focusing time for all analytes is reduced by an order of magnitude because of the order of magnitude reduction in $L_f^2/(e\omega_k V)$. Although separation is faster, dispersion of the focused zones is greater because the Peclet number is 60% lower, resulting in broader, shorter peaks (here $Pe = eV/kT$, which is equivalent to the definition offered with respect to Eq. (11)). The focused analyte peak heights in the

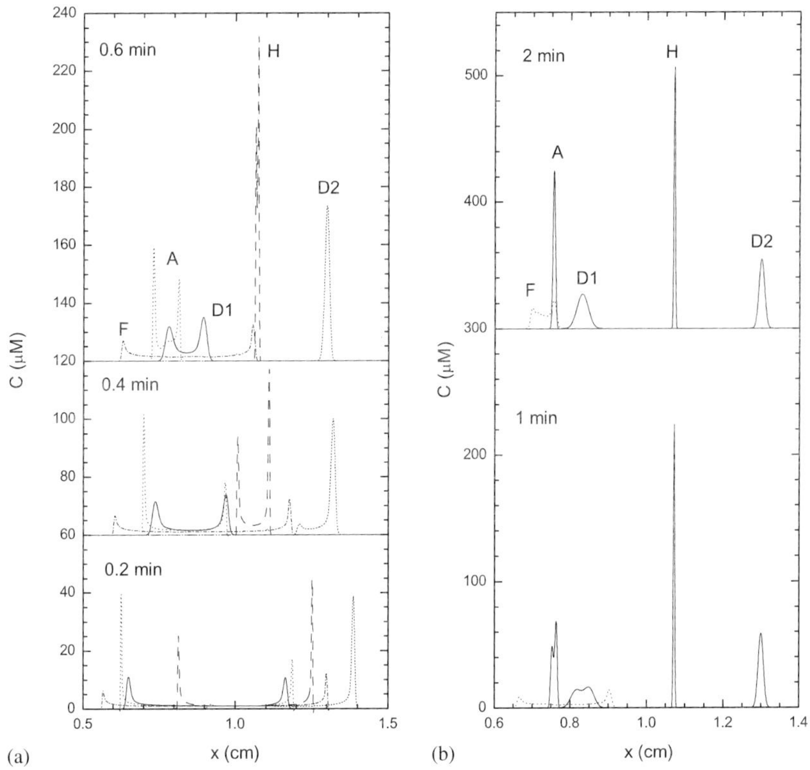

FIGURE 17 Two dyes and three proteins focused between strong acid H_3PO_4 and base NaOH: (a) 0.2–0.6 min; (b) 1–2 min. All analyte concentrations are initially uniform at 1 μM between $x = 0.5$ and 1.5 cm, and zero elsewhere. All other conditions as in Figure 15.

open channel are all about 45% lower than in the closed channel—slightly more dispersed than the 36% predicted by scaling on $\sqrt{Pe}$. There is additional dispersion in the open channel because the electric field peak near the NaOH zone boundary persists (Figure 15), which reduces the field in the remainder of the focusing space. IEF resolution is reduced slightly in open-channel electrolyte configurations because of the migration of the catholyte.

IV. SUMMARY

IEF is an electrophoretic process that hinges on the mobility versus pH behavior of ampholytes. The basic theory of IEF tells us that separands focus about their pI in a Gaussian distribution. Peak height and variance depend on the competition between diffusion away from the pI and electromigration towards it. Comprehensive simulations of IEF involve the numerical solution to conservation laws for all the relevant amphoteric and ionogenic species in the separations milieu. The conservation relations are non-linearly coupled to the driving electric field and through the complicated mass-action relations. With the increased speed of microprocessors, one-dimensional simulations of the type shown here are now readily managed. The simulations offer the possibility of gaining insight into the detailed dynamics of the unfolding separations process. To see a more comprehensive exposition of IEF simulation results, particularly with regard to pH gradient development, the reader should refer the monograph by Mosher et al.[8]

REFERENCES

1. Kolin, A.*J.* Separation and concentration of proteins in a pH field combined with an electric field. *Chem. Phys.* 22:1628, 1954.
2. Svensson, H. Isoelectric fractionation, analysis and characterization of ampholytes in natural pH gradients. 1. Differential equation of solute concentrations at a steady state and its solution for simple cases. *Acta Chem. Scan.* 15:325, 1961.
3. Svensson, H. Isoelectric fractionation, analysis and characterization of ampholytes in natural pH gradients. 2. Buffering capacity and conductance of isoionic ampholytes. *Acta Chem. Scan.* 16:456, 1962.
4. Svensson, H. Isoelectric fractionation, analysis and characterization of ampholytes in natural pH gradients. 3. Description of apparatus for electrolysis in columns stabilized by density gradients and direct determination of isoelectric points. *Arch. Biochem. Biophys.* (Suppl. 1):132, 1962.
5. Vesterberg, O. and Svensson, H. Isoelectric fractionation, analysis and characterization of ampholytes in natural pH gradients. 4. Further studies on resolving power in connection with separation of myoglobins. *Acta Chem. Scan.* 20:820, 1966.
6. Vesterberg, O. Synthesis and isoelectric fractionation of carrier ampholytes. *Acta Chem. Scan.* 23:2653, 1969.
7. Rilbe, H. Historical and theoretical aspects of isoelectric focusing. in *Isoelectric Focusing and Isotachophoresis,* (Catsimpoolas, N. Ed.) *Ann. N.Y. Acad. Sci.*, New York, NY, Vol. 209, pp. 11–22, 1973.
8. Mosher R. A., Saville D. A. and Thormann, W. *The Dynamics of Electrophoresis.* VCH, Weinheim, 1992.
9. Righetti, P. G. *Immobilized pH Gradients, Theory and Methodology.* Elsevier, Amsterdam, 1990.
10. Boris, J. P. and Book, D. L. Flux-corrected transport — minimum-error finite-difference method for solving fluid equations. *J. Comp. Phys.* 18:608, 1973.
11. Anderson, D. A. *Computational Fluid Mechanics and Heat Transfer.* Hemisphere Publ. Co., New York, 1984.
12. Heinrich, J. C. and Pepper, D. W. *Intermediate Finite Element Method.* Taylor and Francis, Philadelphia, PA, 1999.
13. Sounart, T. L. and Baygents, J. C. Simulation of electrophoretic separations by the flux-corrected transport method. *J. Chromatogr. A.* 890:321, 2000.
14. Bier, M., Palusinski, O. A., Mosher R. A. and Saville, D. A. Electrophoresis mathematical modeling and computer simulation. *Science.* 219:1281, 1983.
15. Saville D. A. and Palusinski, O. A. Theory of electrophoretic separations. Part I: formulation of a mathematical model. *AIChE J.* 32:207, 1986.

16. Palusinski, O. A., Graham, A., Mosher, R. A., Bier, M. and Saville, D. A. Theory of electrophoretic separations. Part II: construction of a numerical simulation scheme and its applications. *AIChE J.* 32:215, 1986.
17. Sounart T. L. and Baygents, J. C. Simulatiion of electrophoretic separation: effect of numerical and molecular diffusion on pH calculations in poorly buffered systems. *Electrophoresis* 21:2287, 2000.
18. Henry, D. C. The cataphoresis of suspended particles. Part I. The equation of cataphoresis. *Proc. Roy. Soc. London Ser. A.* 133:106, 1933.
19. Sounart, T. L. and Baygents, J. C. Electrically-driven fluid motion in channels with streamwise gradients in the electrical conductivity. *Colloids Surf. A: Physicochem. Eng. Aspects* 195:59, 2001.
20. Mao, Q., Pawliszyn, J. and Thormann, W. Dynamics of capillary isoelectric focusing in the absence of fluid flow: high-resolution computer simulation and experimental validation with whole column imaging. *Anal. Chem.* 72:5493, 2000.
21. Schmitt, Ph., Poiger, T., Simon, R., Freitag, D., Kettrup, A. and Garrison, A. W. Simultaneous determination of ionization constants and isoelectric points of 12 hydroxy-s-triazines by capillary zone electrophoresis and capillary isoelectric focusing. *Anal. Chem.* 69:2559, 1997.
22. Thormann, W., Molteni, S., Stoffel, E., Mosher R. A. and Chmelik, J. Computer modeling and experimental validation of the dynamics of capillary isoelectric focusing with electroosmotic zone displacement. *Anal. Meth. and Instrum.* 1:177, 1993.
23. Thormann, W., Zhang, C.-X., Caslavska, J., Gebauer P. and Mosher, R. A. Modeling of the impact of ionic strength on the electroosmotic flow in capillary electrophoresis with uniform and discontinuous buffer systems. *Anal. Chem.* 70:549, 1998.
24. Mosher, R. A. and Thormann, W. High-resolution computer simulation of the dynamics of isoelectric focusing using carrier ampholytes: the post-separation stabilizing phase revisited. Electrophoresis 23:1803, 2002.

4 GENERATION OF pH GRADIENTS

TOM BERKELMAN

Life Science Group, Bio-Rad Laboratories, 6000 James Watson Drive, Hercules, CA 94547

I. INTRODUCTION

Isoelectric focusing (IEF) is the electrophoretic separation of proteins in a pH gradient. The pH gradient is therefore critical to the technique and the nature of the pH gradient largely determines the quality and usefulness of the separation. Since its early inception, the technique of IEF has seen improvements in resolution, reproducibility, speed, capacity, and overall applicability, most of which have been made possible by refinements in pH gradient generation. The different IEF methods that have been developed over the years are largely distinguished from one another in terms of the means which by the pH gradient is formed.

The pH gradient should fulfill a number of requirements to be useful for IEF separations. The pH gradient should be reproducible so that the results of separations can be compared with one another. This means that it should be stable with respect to time and other external factors. The shape, slope, and range of the pH gradient should be controllable, so that a pH gradient appropriate to the separation can be generated. The pH gradient should be continuous, as this largely determines the resolution of the separation. Finally, the pH gradient should exist in a medium that

Handbook of Isoelectric Focusing and Proteomics
D. Garfin and S. Ahuja, editors.

provides sufficient conductivity for electrophoretic separation and that ensures the solubility and stability of the proteins to be separated. Other factors may be important in certain situations and are determined by the scale of the separation, whether the separation is preparative or analytical, and how the results of the separation are to be analyzed.

The pH gradient used for IEF can either be "artificial" or "natural." Svensson, an early developer of the technique of IEF, first articulated this distinction,[1] which has subsequently proved useful in understanding the different types of pH gradients used for IEF. An "artificial" pH gradient arises from a gradient in the composition of a solution of acids and bases that is created prior to the application of an electric field. This type of gradient is formed either by the diffusive mixing of a relatively basic solution juxtaposed with a relatively acidic solution, or by the controlled creation of a gradient between two solutions by the use of a gradient mixer. A "natural" pH gradient arises solely through the action of electrical current passing through a solution that is homogeneous prior to application of the current. Establishment of the pH gradient is the result of the same electrochemical processes that drive protein separation. Such gradients have their origin in the electrolysis of water. The solution at the anode becomes acidic due to the electrolytic production of H^+ and the solution at the cathode becomes basic due to the production of OH^-. This results in a pH gradient that can be used for IEF if measures are taken to assure its uniformity, stability, and reproducibility. Both "artificial" and "natural" pH gradients depend on buffering compounds for the maintenance of the gradient and as charge carriers.

II. pH GRADIENTS IN THE EARLY HISTORY OF IEF

IEF can be said to have had its inception in 1912, when Ikeda and Suzuki[2] patented a method for the purification of glutamic acid from a protein hydrolysate. The hydrolysate was placed in a cell comprising three compartments separated by permeable membranes. When an electric current was passed through the cell, it was observed that acidic amino acids accumulated in the compartment closest to the anode, basic amino acids in the compartment closet to the cathode, and neutral amino acids in the middle compartment.

Over subsequent decades, this general method was applied to protein hydrolysates and other samples. These experiments used multichamber electrolyzers in which permeable membranes or siphons separated the chambers from each another. When a sample mixture was introduced into such a device, and a current run through it, amphoteric constituents of the sample separated and accumulated in individual compartments in increasing order of isoelectric point from anode to cathode. This method was eventually extended from the separation of amino acids and other amphoteric small molecules to separations of peptides and proteins. Reports on this general method are numerous and covered comprehensively in a review by Svensson.[3] Although these separations were not IEF in the modern sense, they were undertaken with an understanding of the general principle that proteins and other ampholytes will move under the influence of an electric field until they arrive at the pH zone, where they carry no net charge.[4]

In these early experiments, the pH gradient was "natural" in that it arose solely through the action of electrical current on water. The pH in any intermediate chamber was a consequence of the buffering power that was imparted by the ampholyte accumulating in that chamber. These experiments thus foreshadowed the use of carrier ampholytes to generate pH gradients, but provided neither the resolution nor the reproducibility of later technical refinements.

Early examples of "artificial" pH gradients appeared in the 1950s with descriptions of electrophoretic separations of proteins in applied gradients of buffer either in density gradients[5–7] or on paper strips.[8] These pH gradients were prepared using two different buffers and through simple diffusion. The time scale of these separations was rapid (of the order of minutes), and it was noted that the separation patterns were not stable over time. This was no doubt a consequence of the fact that the buffers used to prepare the pH gradient were themselves electrically charged and mobile within the electric field used to drive the separation. In fact, any pH gradient prepared using charged buffers suffers from this problem. Subsequent developments in isoelectric focusing were largely the result of efforts to overcome this fundamental limitation.

III. THE DEVELOPMENT OF CARRIER AMPHOLYTES

The development of synthetic carrier ampholytes was the first breakthrough that allowed the formation of pH gradients fulfilling everything required to be used routinely for highly resolving IEF. Their development depended in turn on the elucidation of a comprehensive theory of ampholyte-generated pH gradients that pre-dated the introduction of synthetic carrier ampholytes by several years. Svensson presented this theory in a series of papers published in the early 1960s.[1,9,10] This work was the first to introduce the term "carrier ampholyte." The term ampholyte refers to any molecule with a defined isoelectric point resulting from the presence of both basic and acidic functional groups. The term "carrier ampholyte" distinguishes the ampholytes whose purpose is in generating and maintaining the pH gradient from the ampholytes (proteins) that are to be separated by IEF. They are called "carrier ampholytes" because they function to carry charge and buffering power.

Svensson recognized that the "artificial" pH gradients used at the time for protein separations were severely limited by their instability and that proteins can never be expected to reach stable focusing positions in such a gradient. He demonstrated theoretically that a series of ampholytes with isoelectric points distributed over the pH range of interest would form a stable and continuous pH gradient in an electric field. When an electric field is applied across a solution of ampholytes, the ampholytes acquire a positive charge as they approach the anode and a negative charge as they approach the cathode. The charged ampholytes are repelled from the electrodes and at some point in the middle of the electric field they lose their charge and stop migrating. The final position of each individual ampholyte is determined by its isoelectric point. The more basic ones accumulate toward the cathode and the more acidic ones accumulate toward the anode. The final shape of the pH gradient depends on the isoelectric points of the ampholytes, their concentrations, and buffering properties.

The work of Svensson describes, in mathematical terms, the steady state between diffusion and focusing and demonstrates that such pH gradients are stable in an electric field. It is demonstrated that the steady-state concentration distribution of ampholytes in an electric field is a series of overlapping Gaussian curves, one for each ampholyte, ascending in order by isoelectric point from anode to cathode. It is also shown that the complete separation of two proteins by IEF requires the presence of at least one carrier ampholyte with a pI value intermediate between those of the proteins to be separated. This finding has important implications for the design of synthetic carrier ampholyte mixtures. The resolution of the separation depends on the number of different carrier ampholyte species present. Carrier ampholyte mixtures should therefore contain as many different ampholyte species as possible.

Svensson described the properties that a good carrier ampholyte should possess. A stable pH gradient requires high buffer capacity, and effective separation requires high conductivity. Both conditions are satisfied by ampholytes that buffer well at their isoelectric points. A good ampholyte therefore contains at least one basic or acidic group with a pK_a close to its isoelectric point. Ideally, the pI of the ampholyte should be intermediate between two closely spaced pK_a values. Carrier ampholytes should also be small molecules so that they may be separated from proteins following separation, and they should have little or no optical absorbance at wavelengths where proteins typically absorb. A few such compounds were available at the time the study was published, but their number was limited and there were large portions of the pH continuum for which few good ampholytes existed. In the pH range 4–7, no available ampholytes could be identified for which the difference between the pI and the closest pK_a value was less than 1.[9]

IEF in ampholyte-generated pH gradients was first attempted using peptides derived from protein hydrolysates as the carrier ampholyte mixture.[10] These mixtures had many of the properties that a carrier ampholyte mixture should have: they contained a large number of individual ampholyte species, many of which buffered well close to their pI. They also shared many of the properties of the proteins to be separated and could not easily be distinguished from the components of the sample. The principle of carrier ampholyte IEF (CA-IEF) could be demonstrated using peptide mixtures, but this did not lead to a useful analytical or preparative procedure.

Therefore, there was a need for synthetic carrier ampholyte mixtures with the required properties, and this need had to be met before CA-IEF could become a robust and widely practiced technique. The synthetic problem at hand was an unconventional one from the perspective of a synthetic organic chemist. Whereas a synthetic procedure is usually designed to produce a pure compound in as high a yield as possible, carrier ampholytes need to be prepared with a procedure that results in as complex a mixture as possible.

Such a synthesis was in fact developed shortly after the properties of an ideal carrier ampholyte mixture were described. In 1969, Versterberg[11] described the bulk synthesis of a complex carrier ampholyte mixture that satisfied the criteria outlined by Svensson. A mixture of diamines and polyamines are reacted with acrylic acid. The amines condense with the unsaturated bond of the acrylic acid by Michael addition and the result is a complex mixture of polyamino polycarboxylic acids. Figure 1 shows the basic chemical scheme. The dissociation constants of the amines are distributed over a pK_a range from 3 to 10. Varying numbers of carboxylic acid groups per ampholyte molecule balance the basic amine groups, resulting in a range of pI values also covering the pH range from 3 to 10. The pK_a values of the amines for any given ampholyte are closely spaced, resulting in the high buffer capacity and conductivity predicted to be necessary for the generation of a stable pH gradient useful for IEF.[9] These

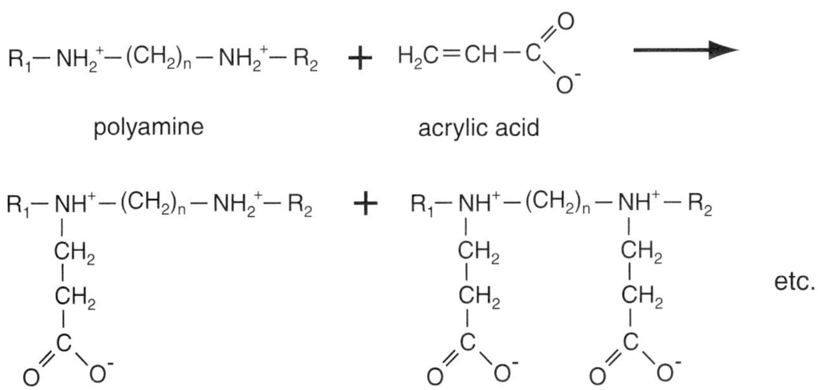

FIGURE 1 Chemical scheme for the synthesis of Ampholine™ carrier ampholytes.

carrier ampholytes also have the desirable properties of low optical absorbance at 280 nm and an average molecular weight well below that of most proteins. They are also very soluble in water. Although the synthetic procedure gives a carrier ampholyte mixture a broad pI range, narrower pH ranges can be prepared by isoelectric fractionation of the parent mixture.

A patent was issued for this synthetic procedure[12] and the carrier ampholyte mixture was produced at various pH ranges and distributed by LKB Produkter, AB (now Amersham Biosciences) under the trade name Ampholine™.[13]

This basic procedure was in time adopted, with variations, by other manufacturers in the production of their own carrier ampholyte mixtures (e.g., Bio-Lyte™ from Bio-Rad, Resolyte™ from BDH), most of which were similar to Ampholine™ in their general properties. There were also publications of straightforward syntheses of carrier ampholytes that utilized commercially available precursors in a synthetic scheme accessible to the laboratory researcher.[14,15] All start with a mixture of polyamines that contains a wide distribution of pK_a values, which is treated to convert primary and secondary amines into tertiary amines with the addition of acidic functionality.

One noteworthy variant of the basic Vesterberg scheme was described and patented by Grubhofer[16,17] and distributed by Serva Feinchemie GmbH under the brand name Servalyt™. This method starts with the usual polyamine mixture, and acidic functionality is provided by sulfonic and phosphonic acid moieties. Sulfonate groups are introduced into the ampholyte mixture through the reaction of propane sultone with amine, and phosphonate groups are added through the reaction of chloromethylene phosphonic acid with amine. The basic chemical scheme for this synthesis is shown in Figure 2.

Although the primary purpose for the development of this synthesis was no doubt the circumvention of the Vesterberg patent,[12] this carrier ampholyte mixture has some attractive properties relative to the LKB product. Additional acidic functional groups give a more continuous and broader pH gradient, and the sulfonic and phosphonic acid groups in Servalyt™ do not chelate metal ions as has been observed in the case of Ampholine™.[13]

The one exception to the general scheme of adding acidic groups to polyamines came with the development of Pharmalyte™ carrier ampholytes, first patented and distributed by Pharmacia Fine Chemicals AB[18,19] (now Amersham Biosciences). This mixture was prepared by the co-polymerization of amines and amino acids with epichlorohydrin. In contrast with the syntheses described above,[11–17] in which a mixture of acids and bases containing a wide range of pK_a values produces a mixture with a wide pI range, these carrier ampholytes are synthesized from starting materials that have a relatively narrow range of pK_a values. This

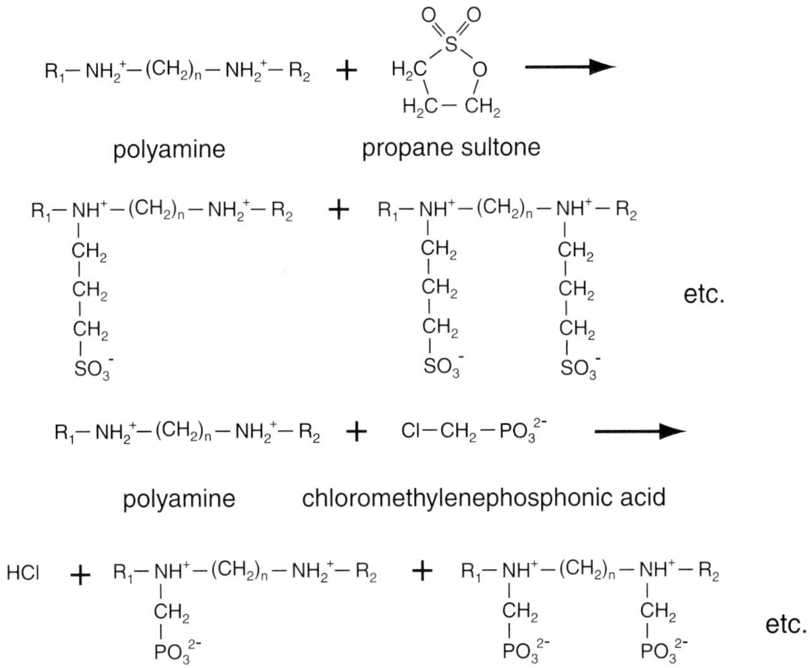

FIGURE 2 **Chemical schemes for the synthesis of Servalyt™ carrier ampholytes.**

assures good buffering capacity and yields carrier ampholytes with relatively narrow pH ranges (typically only about 2 pH units). Wider pH ranges are prepared by mixing together different narrow range mixtures. The manufacturer claims that this mixture contains more individual ampholyte species than are produced with the standard method and therefore provides a smoother, more highly resolving pH gradient.

IV. PRACTICAL ASPECTS OF CARRIER AMPHOLYTE-GENERATED pH GRADIENTS

The commercial availability of carrier ampholyte mixtures put the robust and straightforward separation technique of CA-IEF in the hands of researchers, and a number of near simultaneous reports of IEF separations were published as soon as Ampholine™ appeared in the market.[20–26] The practice of CA-IEF has many variations, most of which concern how the pH gradient is stabilized and whether the separation is analytical or preparative.

Common features of the various CA-IEF methods include the use of carrier ampholytes at a concentration of 2–4% (w/v) and the use of anolyte and catholyte reservoirs that are typically at or beyond the pH extremes of the gradient. The use of anolyte and catholyte stabilizes the pH gradient by providing end points and restricting the ampholytes to the region where protein separation is to occur. Resolution of the technique depends on the electric field strength, so the technique is usually performed using instrumentation capable of delivering field strengths of 50 V/cm or higher. High field strengths result in Joule heating, so instrumentation that provides effective heat dissipation is essential.

A means of gradient stabilization is required for CA-IEF because convective mixing otherwise abolishes any pH gradient. Carrier ampholyte-generated pH gradients have conductivity variations along the gradient and, consequently, variations along the gradient in the amount of Joule heat generated during IEF. It is this non-homogeneous heat generation that can result in convective mixing. The earliest reported CA-IEF separations were performed in columns in which the pH gradient was stabilized with a sucrose density gradient.[27,28] A sucrose gradient containing a homogeneous mixture of carrier ampholytes is poured into a glass column with the aid of a gradient mixer. Voltage is applied across the length of the column in a device that allows the venting of electrolysis gases that would otherwise disrupt the gradient. Many proteins will focus sharply under these conditions, and the relative simplicity of collecting fractions from such a device brought this technique early prominence as a preparative method. However, as proteins focus and concentrate, they form zones of higher density that destabilize the density gradient. Moreover, many proteins aggregate and precipitate as they focus. When this occurs, they sediment out of the gradient. These problems, as well as the difficulty of manufacturing the instrument, resulted in the decline of this technique. Instruments for preparative density gradient column IEF are no longer commercially available and the technique is rarely practiced.

Other preparative modes of CA-IEF are, however, still widely used. The Rotofor™ instrument (Figure 3), available from Bio-Rad Laboratories, uses rotational stabilization. Rotating the column in which the separation is conducted prevents convective mixing and settling.[29,30] The separation column is divided into individual chambers by permeable membranes, and fractions are collected following separation by simultaneously evacuating all of the chambers. IEF is also performed in beds stabilized by granulated matrices such as Bio-Gel™ P and Sephadex™.[31,32] Collection of material following separation involves removing the medium from the zone of interest and eluting the protein from the matrix.

Analytical CA-IEF was originally performed using polyacrylamide gel as the stabilizing matrix,[20–26,33–36] and this remains the dominant method for analytical CA-IEF. Polyacrylamide gels are easily prepared, and the resulting matrix is transparent and largely free of fixed charge. The gels can be cast as tubes or as slabs, which can be run either vertically or horizontally (Figure 3). Separated proteins can be visualized by staining; most commonly used staining methods (e.g., silver, Coomassie™

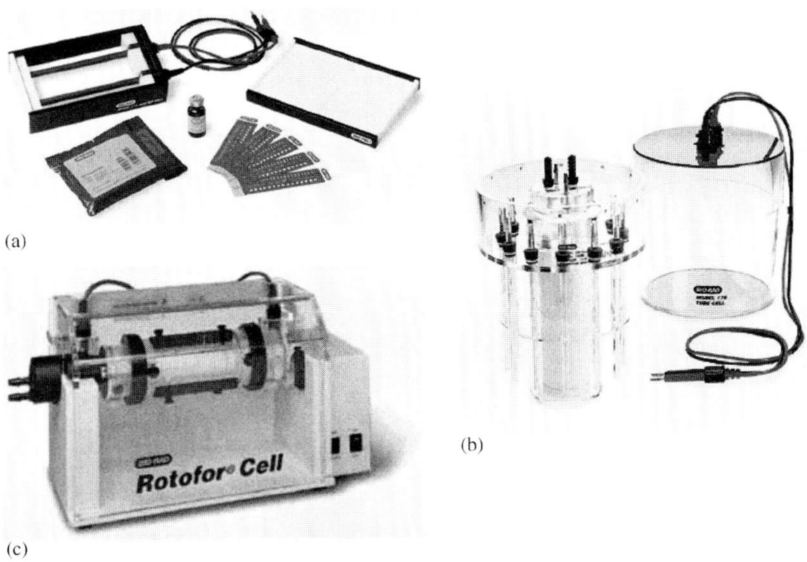

FIGURE 3 Instrumentation used for various modes of CA-IEF: (a) Bio-Rad model 111 mini IEF cell for horizontal electrophoresis of IEF slab gels. (b) Bio-Rad model 175 tube cell for vertical electrophoresis of IEF tube gels. (c) Bio-Rad Rotofor cell for preparative CA-IEF with rotational stabilization. From Bio-Rad Laboratories with kind permission.

Blue, fluorescent stains) can be applied to CA-IEF gels with some modification.

Agarose gels are also widely used for analytical CA-IEF,[25,37] as the larger pore size of agarose matrices allows the separation of larger proteins not amenable to separation in polyacrylamide.[38] This technique became particularly useful after methods developed for the production of uncharged agarose allowed electrophoresis in this medium without electroosmotic flow.[39,40]

CA-IEF is also performed in capillaries.[41–43] The high voltage delivered by capillary electrophoresis instruments, along with the very effective heat transfer within narrow capillaries, make this an exceptionally highly resolving and rapid separation technique.

CA-IEF separation patterns are sharper and more reproducible in the presence of urea at 8 M or higher. This effect is mostly due to the positive solubilizing effect that this denaturant has on proteins. Other factors may include the stabilizing effect of high viscosity and the disruption of interactions between different ampholytes. This fact has been exploited in a 2-D electrophoresis method that has probably been the most widely used application of CA-IEF.[44] In the first dimension of this technique, a complex protein sample is applied to a polyacrylamide tube gel that is cast with carrier ampholyte, urea, and non-ionic detergent. Following separation, the tube gel is equilibrated in SDS-containing buffer and applied to the top of a slab gel for size-based SDS–PAGE separation.

As should be expected, carrier ampholytes produced by different suppliers by different methods differ in their physical properties and performance in isoelectric focusing. A number of comparative studies have been published comparing molecular weight distributions, buffer capacity, and resolution,[45–50] but a practical selection of the optimal carrier ampholyte mixture to be used in a given situation is best determined empirically. Buffer capacity and conductivity of the different mixtures is found to vary slightly across the pH spectrum, so different mixtures tend to perform better in different pH ranges.[45,46] The average molecular weight of a carrier ampholyte mixture can be a consideration during 2-D electrophoresis, as carrier ampholytes can co-migrate with smaller proteins during the second dimension separation and interfere with protein visualization. The size distribution of carrier ampholytes is also found to vary across the pH spectrum, but Pharmalyte™ has generally been found to have a higher average molecular weight than any of the carrier ampholyte mixtures prepared by the method of Versterberg.[45–47]

Given Svensson's finding that the resolution of CA-IEF can depend on the number of individual carrier ampholyte species present in a mixture, it would seem that there might be an advantage to mixing carrier ampholytes from different suppliers in order to maximize the number of individual species. This practice does not seem to be widely applied, but carrier ampholytes of different ranges are often mixed in order to customize the shape and slope of the pH gradient generated. As an example, carrier ampholytes with a pH range of 5–8 can be blended with carrier ampholytes with a pH range of 3–10 to generate a pH gradient from 3 to 10 in which the slope of the gradient in the range from 5 to 8 is flattened in order to maximize the resolution of proteins in that pH range.

V. LIMITATIONS OF THE CARRIER AMPHOLYTE METHOD

Although CA-IEF provides protein separations of unparalleled resolution, limitations of the technique became apparent soon after its introduction. Carrier ampholytes, being a complex and poorly characterized mixture, are not always synthesized reproducibly, and variations among the pH gradients produced by different batches can occur.[51] Discontinuities in carrier ampholyte-generated pH gradients have also been reported.[52] There appear to be fundamental limitations to the pH ranges possible with CA-IEF. Attempts to synthesize carrier ampholyte mixtures covering ranges falling outside of the pH range of 3 to 10 have been unsuccessful,[48] and the width of an effective carrier ampholyte-generated pH gradient has a practical lower limit of about 1 pH unit.

Carrier ampholyte-generated pH gradients are also unstable over time.[53–59] Problems include a progressive flattening of the gradient and a loss of the basic end of the gradient (termed "cathodic drift").[56,57] Much has been published about cathodic drift and its causes. It has been attributed variously to electroendosmosis,[60] absorption of CO_2,[61] slow electrophoretic migration of ampholytes,[59,62] aggregation of ampholytes[63] and diffusion of ampholytes out of the gradient.[64] Given the diversity of opinion on the subject, it is likely that cathodic drift is complex in origin and may have different causes in different situations.

Cathodic drift can be mitigated slightly by the addition of viscosity agents such as urea or glycerol,[65] the adjustment of anolyte and catholyte pH,[62] and performing IEF in a CO_2-free atmosphere.[61] There is, however, no comprehensive solution to the problem. As a result of cathodic drift, CA-IEF in the range above pH 8; this pH range is very difficult in practice.[65,66] Since many proteins have isoelectric points in this range, this has remained a fundamental limitation of CA-IEF.

One method that has been developed to circumvent the problem of cathodic drift and allow the separation of basic proteins is the technique of non-equilibrium pH gradient electrophoresis (NEPHGE).[67] In this technique, the protein sample is applied to the anodic end of a carrier ampholyte-containing gel and current is applied in the same manner as for conventional CA-IEF. In NEPHGE, however, the run is terminated before the proteins have reached their equilibrium, focused positions. In this manner, proteins with isoelectric points above the range specified by the carrier ampholyte mixture remain within the gel. This technique is empirically observed to provide good resolution of basic proteins, although the physical mechanisms allowing proteins to be well resolved under non-equilibrium conditions are poorly understood. This technique has been used in two-dimensional separation of basic proteins, but has been criticized for its lack of reproducibility.[68]

The preparative use of IEF with carrier ampholytes is limited by the difficulty of removing carrier ampholyte from purified protein preparations. Downstream applications of the purified protein often require a preparation free of carrier ampholyte contamination, yet the similar physical properties of protein and carrier ampholyte render the separation difficult. Carrier ampholytes have also been found to form associations with proteins that can be difficult to disrupt.[69,70] Methods of carrier ampholyte removal have been published and include dialysis, ammonium sulfate precipitation, and ion exchange.[48,71–74] These procedures are, however, rather complicated and time-consuming.

VI. EARLY ALTERNATIVE IEF MODES NOT REQUIRING CARRIER AMPHOLYTES

The above-mentioned limitations of carrier ampholytes, and the expense of the larger quantities of carrier ampholytes required for preparative applications prompted the development of a number of methods for generating pH gradients for IEF without the use of carrier ampholytes. Although none of these methods was ever widely adopted, they are of interest from a historical perspective.

Since the pK_a of the buffer varies with temperature, a temperature gradient imposed across a constant buffer concentration will result in a pH gradient. The effect is particularly pronounced with Tris buffer. This phenomenon has been exploited to generate pH gradients for IEF and is capable of creating pH gradients up to a single pH unit in width.[75,76] The pK_a of the buffer can also depend on the dielectric properties of the solution, and a method has been developed for isoelectric focusing in pH gradients established using a gradient of a water-miscible, high-dielectric organic component.[77] This method can yield gradients of several pH units. In both these cases, the negative effect of elevated temperature or organic solvent on protein stability and solubility probably limited the widespread use of these techniques.

Rilbe (formerly Svensson) conceptualized a method for performing isoelectric focusing in simple buffers. Termed "steady-state rheoelectrolysis," this method aimed to create a stable pH gradient in an unmixed zone of buffer solution between anodic and cathodic compartments delimited by permeable membranes.[78] Normally, when a simple buffer solution is electrolyzed in such an apparatus, the acid is drawn to the anode and the base is drawn to the cathode. The result in the absence of mixing is a sharp discontinuity from low pH to high pH in the middle of the apparatus, rather than a continuous pH gradient. Rilbe proposed an instrument that continually pumped the base back to the anode and the acid back to the cathode. He postulated that a continuous pH gradient could be achieved in this manner. Some success in generating pH gradients was achieved with prototype instruments, but the method was never put into practice for separating proteins.

More success was achieved in systems using two different buffers. Bier et al. demonstrated that narrow (up to 1 pH unit) but stable pH gradients could be generated with pairs of simple buffers. If both the buffers are selected to be over 90% non-ionized over the pH range of the separation, there is little buffer movement over the course of the separation. The resultant pH gradient was indeed found to be remarkably stable.[79] For a time, Bio-Rad Laboratories commercialized this method for use with Rotofor™ apparatus as a series of buffer pairs termed RotoLyte™ buffers. There are a number of advantages to this approach: buffer concentrations of up to 100 mM can be used, which allows the separation of proteins that require relatively high ionic strength for solubility. The components of the system are well characterized, show little tendency to interact with proteins, and are easily separated from the purified proteins following separation. The technique also allows highly resolving preparative separations in very narrow pH gradients.

VII. IMMOBILIZED pH GRADIENTS

Of all the pH gradient generation schemes developed in response to the disadvantages of carrier ampholytes, the most successful and widely applied has been the so-called immobilized pH gradient (IPG). This method is, in some sense, a return to the early "artificial" pH gradients generated from applied gradients of non-amphoteric buffers.[5–8] The early methods were ultimately impractical because the buffers responsible for the pH gradient were themselves mobile in an electric field, thus rendering the pH gradient unstable. In an immobilized pH gradient, the buffers are covalently grafted to a support or matrix, preventing their movement. This results in a truly stable "artificial" pH gradient.

The IPG principle was first described in a patent by Rosengren et al., which described the preparation of immobilized pH gradients by the covalent incorporation of buffers into a gel matrix by co-polymerization.[80] The first IEF separations using immobilized pH gradients were presented by Bjellqvist et al. in a seminal report published in 1982.[81] The report describes a series of bifunctional buffers, each consisting of an acidic or basic buffering group connected to an acrylamido moiety via linkage to the amide nitrogen. These compounds are of the general structure CH_2=CH–CONH–R, where R is the buffering group. Synthesis of these compounds is straightforward and is accomplished by reacting acryloyl chloride with a suitable amine derivative. In the case of acidic buffers, an amino acid is used. In the case of basic buffers a diamine is used. This diamine contains a primary amine that reacts with the acid chloride and a tertiary amine that acts as the buffering species.

The reactive vinyl groups on these molecules allow them to be copolymerized into a polyacrylamide matrix to generate immobilized pH gradients. Linear pH gradients were prepared using a gradient mixer and two mixtures of acrylamido buffers: one relatively basic and one relatively acidic. The buffer mixtures in this first report were relatively simple, consisting of only two or three different acrylamido buffers. As a result, only narrow pH gradients were possible, spanning at most two pH units.[81]

The immobilized pH gradients initially reported utilized a series of six acrylamido buffers, two acids ($pK_a = 3.6$ and 4.6), and four bases ($pK_a = 6.2$, 7.0, 8.5, and 9.3). These were produced by LKB Produkter AB (now Amersham Biosciences) and sold under the trade name Immobiline™. Additional acrylamido buffers, in particular a basic acrylamido buffer with a pK_a of 10.3 as well as a strongly acidic sulfonic acid derivative and a strongly basic quaternary amine derivative, also came into use.[82] Structures of the acrylamido buffers used for immobilized pH gradients are shown in Figure 4.

Immobilized pH gradient IEF (IPG-IEF) has several advantages over CA-IEF, which were apparent even at its inception. Immobilization of the buffering species essentially conquers the problems of gradient flattening stabilized focused bands against precipitation. These gradients tended to be physically unstable and required elution from the bottom of the columns for protein detection, both of which resulted in loss of resolution.

The first attempt at employing more stable media and more compact geometry was that of Wrigley,[2] who used disc electrophoresis equipment to run IEF separations in cylindrical polyacrylamide gel rods in glass tubes. Here, too, pH gradients were established with carrier ampholytes. Because staining of the tube gels in the presence of carrier ampholytes was problematic, two-dimensional separation technique was developed, which combined the gel-rod technique with thin-layer electrophoresis.[3]

The cylindrical geometry of both sucrose gradients and gel rods allow sample loading only at one end of the gradient, i.e., either at the anode or at the cathode side. It was soon recognized by Awdeh et al.[4] as well as Leaback and Rutter[5] that horizontally placed slab gels would allow sample loading anywhere along the pH gradient, and, moreover, such gel layers would allow direct comparison of different samples in the same gel.

Figure 1 shows the first apparatus for performing IEF in polyacrylamide slab gels.[4] In this apparatus, two carbon electrodes are mounted 20 cm apart in a simple plastic box. A glass plate holding the gel rests horizontally on the carbon electrodes that are connected to an external power supply. The gel is cast between two glass plates separated by a 1-mm-thick silicone rubber gasket and clamped together with spring

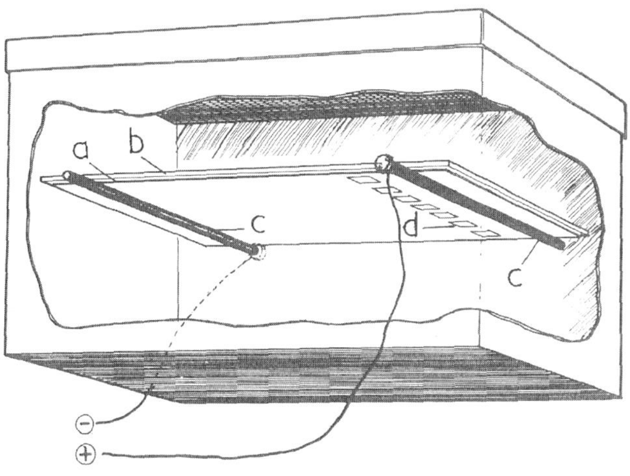

FIGURE 1 The first apparatus for slab gel IEF in polyacrylamide matrix according to Awdeh, Williamson, and Askonas. (a) polyacrylamide gel layer 1 mm thick; (b) glass plate; (c) carbon electrodes 20 cm apart; (d) site of sample application. Each sample, 100/400 μg of protein in less than 50 μL, is pipetted onto the surface of the gel and spread over a rectangular area about 1 × 2 cm (reproduced after Awdeh et al.[4]).

clips. The monomer solution, mixed with carrier ampholytes, is pipetted into this cassette and left to polymerize overnight. Coating one of the glass plates with a thin layer of silicone grease makes it easy to remove from the polymerized gel surface while the gel slab adheres to the other plate. Samples are applied by first pipetting them onto small filter paper squares and then inserting the strips into short slots cut at the desired application points. For the separation, the carbon cathode is often moistened with 5% (v/v) ethylene diamine and the anode with 5% (v/v) phosphoric acid, but this is not always necessary. The glass plate holding the gel and sample papers is placed inverted onto the carbon electrodes. In order to prevent drying of the gel, the walls of the box are sometimes covered with plastic sponge sheets moistened with water. The lid is closed and the box is placed in a chilled environment, and a voltage of 400 V is applied to the electrodes for about 4–16 h. With the original device, active cooling of the gels was attempted by applying a puddle of gasoline on the glass plate and taking advantage of its latent heat of evaporation. Obviously, IEF was a dangerous business at this time.

Some alternative matrix materials for carrier ampholyte IEF have been tried, such as agarose gels, granulated dextran or polyacrylamide beads, and cellulose acetate foils. A comprehensive description of all these matrices can be found in the book by Righetti.[6] This chapter describes the most successful approaches to slab gel IEF.

II. EQUIPMENT

A. Isoelectric Focusing Chamber

In standard IEF chambers, such as those sold by Amersham Biosciences and shown in Figure 2, gels are placed horizontally on water-cooled plates with inert surfaces with the gel surface facing upward. The aluminum ceramic plates of standard apparatus contain cooling coils for even temperature distribution over the gel area and accurate temperature control of the gel. The cooling plates are connected to external thermostatic circulators as shown in Figure 3.

Platinum electrodes are usually used and they are connected to an external power supply. Electrical contact between the gel edges and the electrodes is made with paper strips soaked with electrode solutions: an acid for the anode and a base for the cathode. The lids of well-made chambers are closed tightly to avoid humidity entering the chamber, because the development of water condensation should be kept to a minimum. Some models provide space in the chamber for wide filter strips soaked in NaOH solution to absorb carbon dioxide from the air. Reservoirs of limestone granules sometimes serve the same purpose. This is done because carbon dioxide absorbed into the gel carboxylic acid and shifts the basic

C. Cathode Drift

Finlayson and Chrambach[7] observed a so-called plateau phenomenon in pH gradients that occurs after a certain focusing time: the pH gradient becomes steeper at the two ends and flattens in the center. Independently, Righetti and Drysdale[8] detected the same effect, which they called "cathodic drift." This instability of the pH gradient is more pronounced in the basic area, hence the cathode drifts is more obvious than the plateau phenomenon. It is time and voltage-dependent, and leads to losses of proteins at the two ends of the gradient. Righetti has "collected" seven reasons for cathode drift:[6] (1) electrophoretic migration of carrier ampholytes, (2) electroendosmosis, (3) formation of a pure water zone at neutral pH, (4) IEF of water at pH 7 causing a backflow toward the electrodes, (5) progressive decay of carrier ampholytes, (6) gain or loss of charged ligand of carrier ampholytes, and (7) diffusion of electrolytes into the gel.

In order to keep the cathode drift to a minimum, focusing voltage and time should both be held to the minimum needed for proper focusing the proteins in the sample. Very often the focusing conditions are controlled with the help of the volt–hour integral.

Care must be taken to use inert matrix ingredients. In agarose gels, cathode drift is more pronounced than in polyacrylamide gels. In denaturing IEF (see below), the cathodic drift causes more problems than in native gels, because of the longer focusing times necessary to move the unfolded and bulky proteins through a highly viscous urea–gel matrix

IV. POLYACRYLAMIDE GELS

A. Gel Composition

Polyacrylamide is an ideal matrix for IEF with both carrier ampholytes and immobilized pH gradients. It is chemically inert, compact, completely transparent, and free of electroendosmosis (see above). Moreover, polyacrylamide can be formed in any shape, in particular as rectangular slabs. Polyacrylamide gels are formed in cassettes by free-radical-induced polymerization in aqueous mixtures of acrylamide and with *N*,*N*′-methylenbisacrylamide (Bis) as crosslinker. Polymerization initiators such as tetramethylethylenediamine (TEMED) and ammonium persulfate generate the requisite free radicals for the reaction. The reaction is best performed in oxygen-depleted solution, because oxygen is a potent inhibitor of polymerization. Two glass plates separated by a gasket and held together with clamps form a gel cassette. As already mentioned, in IEF, it is desirable to keep the retardation effect of the matrix as low as possible. Large pore-size gels, usually gels with a total acrylamide concentration of 5% *T* and a crosslinking factor of 3% *C* are used.

The T and C values of the gels are defined as follows:

$$T = \frac{a+b}{V} \times 100(\%),\ C = \frac{b}{a+b} \times 100(\%) \qquad (1)$$

where a is the mass of acrylamide in g, b the mass of methylenebisacrylamide in g, and V the volume in mL.

With less than 5% T and/or 3% C gels are not easily handled, because they become too soft. C values higher than 7% C increase the pore size but lead to very hydrophobic and brittle gel matrices. The reagents need to be of high quality, because the IEF method is very sensitive to contaminants and electroendosmosis (see above).

Gels should be polymerized overnight in order to complete the "silent polymerization" that occurs after gel formation. That is, overnight polymerization is beneficial because it exhausts unreacted monomers. If gels are used too early, interactions between reactive monomer compounds and some proteins can occur.

B. Gel Geometry

Some laboratories use vertical slab gel electrophoresis equipment for IEF. In this case, the gel remains in the casting cassette between two glass plates during the run. For the separation, the cassette is placed vertically between two buffer tanks. The disadvantages of this arrangement are: (1) sample can only be loaded at the end of the gradient; (2) the buffer tanks need to be filled with large volumes of electrode solutions; (3) as IEF separates only according to the charge, the gel matrix must contain large pore sizes. Such a soft gel can slide down between vertical glass plates. The gel should preferably be cast on a film support. (4) The contact to glass surfaces leads to electroendosmosis and thus to enhanced cathodal drift. (5) Temperature in most of these chambers cannot be well controlled. IEF has to be performed under active temperature control. (6) Vertical electrophoresis chambers are not suitable for high voltages. The focusing step needs high electric field strength. Therefore, the horizontal flatbed chambers with cooling devices—as described above—are the best choice.

C. Carrier Ampholyte Gels

It is usual to add 2% (w/v) carrier ampholytes to the monomer solution prior to polymerization. Unless a supplier indicates otherwise, it should be assumed that the dry-weight content of a carrier ampholytes stock solution is 40% (w/v). If less than 2% carrier ampholytes are used, the buffering capacity can be insufficient, whereas if more than 2% are used, the loading capacity is reduced because the carrier ampholytes compete with the proteins for the water in the gel.

1. Choices of Carrier Ampholytes

Carrier ampholytes are mixtures of 600–700 different homologues of amphoteric compounds with a spectrum of isoelectric points between 3 and 11. They form pH gradients under the influence of electric fields. These substances have high buffering capacities at their isoelectric points. They have molecular weights below 1kDa and do not bind to proteins because they are highly hydrophilic. Narrow interval mixtures are available to increase the resolution and for the selection of defined isoelectric point ranges.

The differences between carrier ampholytes of different suppliers are based on the chemistry of their production. The original carrier ampholytes, developed by Vesterberg[9] and marketed under the name "Ampholine™," are produced by reacting aliphatic oligoamines with acrylic acids. Other products are co-polymers of glycine, glycylglycine, amines, and epichlorhydrin. Thus, the isoelectric points and buffering properties of the individual homologs of the different products are slightly different and produce different results. For the sake of reproducibility within an experimental series or for consistency of results in a validated analysis it is important to stay with one product. Some laboratories mix the carrier ampholytes from different suppliers to achieve a mixture with a higher number of different homologs in order to obtain smoother pH gradients.

2. Gel Polymerization

There are two ways to polymerize polyacrylamide gels: (1) chemical polymerization with TEMED and ammonium persulfate; and (2) photopolymerization with riboflavin and TEMED. The choice of the polymerization method is dependent on the composition of the carrier ampholytes, which is different for the various commercial products, and on the pH range. Righetti and Caglio[10] found that polymerization efficiency with the persulfate–TEMED system is most efficient above pH 6, whereas photopolymerization with riboflavin–TEMED is most efficient below pH 6. In the presence of carrier ampholytes it is not necessary to add TEMED, because carrier ampholytes contain primary amino groups to initiate the chain reaction. In this case, photopolymerization has the advantage that no ionic additive is added. In general, chemical polymerization is easier to perform, because no light source is needed.

3. Prepolymerized Gels

If unreacted reagents from polymerization are of concern, gels can be prepolymerized without carrier ampholytes, then washed and dried and rehydratated with carrier ampholyte-containing solution before use.[11] This rehydratation concept works best, of course, with very thin slab gels. Rehydratation can be performed in the same glass cassette used for casting the gel, or in a horizontal plastic tray with a defined volume of rehydratation solution. It is very important that the surface of such a tray is not smooth, because the surface of the incompletely swollen gel could stick to the bottom of the tray. These washed gels can be run without electrode solutions; no filter-paper strips between electrodes and the gel edges are needed.[12] Practice has shown that zymogram techniques (see Section D 2) function much better with washed and rehydratated gels, because the reactive catalysts and non-reacted acrylamide monomers have been removed from these gels.

4. Ultrathin-layer IEF Gels

Soft IEF gels are not easy to handle during fixing and staining, thus gels that were 2–3 mm thick were the first to be used. The first commercial precast gels were 1 mm in thickness and supported on Mylar sheets. For preparing very thin gels in the laboratory, cellophane film can be used to achieve mechanical stability. As shown in Figure 4, thinner gels show sharper bands and higher resolution than 1 mm and thicker gels.[13] The thickness of lab-cast gels is often defined by the number of layers of Parafilm® used as gaskets: one layer is 120 μm thick. It does not help to make gels thinner than 200 μm. When gels with a thinner layer are used, the gradient starts to shrink to a shorter distance, leaving two plateaus without any bands on both ends of the gels (unpublished observations by the author). Radola suggested use of small 50 to 100-μm thin gels with short separation distances.[14] These extremely thin gels with 5 cm edge length are very sensitive to salt or other contaminations in the sample, and can only be used to display highly purified proteins.

In comparison with standard gel thickness, ultrathin gels offer faster separation, higher sensitivity in staining, and cost saving because of the

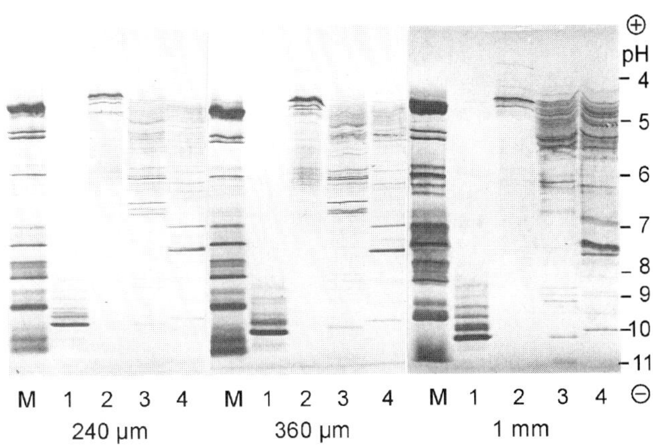

FIGURE 4 **The influence of gel thickness on the resolution in IEF. Coomassie Brilliant Blue R-250 staining. Lane (M) marker proteins: (1) trypsin; (2) soybean lipoxidase; (3) and (4) legume seed proteins (reproduced after Görg et al.[13]).**

lower carrier ampholytes consumption. Ultrathin-layer gels are also very useful for the detection of peptides with specific stains[15] using a printing technique.

5. Additives

Proteins can be focused under native or denaturing conditions, in practice, native conditions are mostly employed. Denatured proteins show only one IEF band, whereas different native conformations can result in multiple-band patterns. Under native conditions proteins are present in multiple conformations, which often results in multiple-band patterns for single proteins. Treating samples with 8 M urea and a reductant, such as dithiothreitol or 2-mercaptoethanol, and using a gel containing 8 M urea, provides denaturing conditions in which the proteins are converted into their constituent polypeptides and the solubility of hydrophobic proteins is considerably increased. Denaturing conditions are mainly chosen for separating highly hydrophobic proteins.

For separations under denaturing conditions the gel must contain 8 M urea.[16,17] The isoelectric points measured in urea gels are different from those in native gels. Therefore, information about the urea content of an IEF run is an important part of the definition of an isoelectric point of a protein. In order to avoid crystallization of urea, gels must be run at 20°C. At temperatures higher than 20°C, formation of isocyanate from breakdown of urea can result in differential carbamylation of proteins leading to artifactual bands. For this reason, temperatures greater than 30°C should be particularly avoided. Carbamylation can also occur, when urea of low purity or old urea solutions are used. Urea IEF gels are mainly employed for the separation of proteins with poor solubility. An interesting example is IEF of caseins as the official method to detect quantitatively how much cows' milk might be illegally used for the production of goats' milk and feta cheese.[18]

As shown by Jenne et al.[19] and by Altland and Hackler,[20] with a denaturing urea gradient perpendicular to the pH gradient in a slab gel protein oligomers and protein–protein-additives can be studied.

When non-ionic or zwitterionic detergents have to be included for solubilization of very hydrophobic proteins, polymerization of gels on support film is not efficient and can result in the separation of gel and foil during IEF. Moreover, handling of these gels is almost impossible. Using dehydrated gels solves the problem.

Reductants cannot be included in gel monomer mixtures, because these reagents inhibit the gel polymerization. If it is necessary or desired to include them during an IEF run, only prepolymerized, dehydrated gels can be used.

For the analysis of temperature-sensitive proteins and protein–enzyme complexes at sub-zero temperatures it is possible to polymerize gels with 37% dimethylformamide[21] to prevent freezing of the matrix.

When samples with different protein concentrations and salt loads are applied on slab gels, the different conductivities in the adjacent lanes can lead to strongly distorted iso-pH lines. Addition of 10% sucrose or 10% ethylene glycol to the polymerization or rehydratation solution reduces this effect and leads to straighter band patterns.

In order to optimize the resolution in certain pH zones, gradients can be modified by adding amphoteric separator compounds to the carrier ampholytes. For instance, adding 0.33M β-alanine to a pH gradient of 6 to 8 and running IEF at 15°C increases the resolution close to the major haemoglobin band HbA in order to detect the glycosylated variant HbA1c in the area around pH 7.15.[22] See Righetti's book[6] for a comprehensive overview on pH gradient modification for the separation of hemoglobin variants.

6. Sample Properties

(a) Salt Effects: In general, IEF runs are sensitive to salt. When samples with different salt concentrations are applied in adjacent lanes of slab gels, the band patterns can show wavy iso-pH lines. This is especially true for gradients generated with carrier ampholytes, so it is best to limit the salt concentration to 50 mM or less. If the salt concentration is too high, the sample should be dialysed against 1% glycine or 1% carrier ampholytes, or desalted with a gel filtration column. For the case that only very small sample amounts are available and dialysis or gel filtration is impractical, adjustment of all samples to the same salt content by diluting them with salt solutions can often solve the problem. In this case, a longer sample entrance phase with low voltage should be applied in order to transport the salt ions into the electrode strips without overheating the gel.

(b) Double One-Dimensional Electrophoresis: In double one-dimensional electrophoresis,[19,20] two different types of slab gel electrophoretic separations are combined as described in the following example. Multiple samples of human serum sample were first fractionated by IEF in a slab gel. A strip containing the IgG fractions was cut out across all sample lanes and transferred onto the stacking gel surface of a flatbed discontinuous electrophoresis gel in order to separate the low-molecular-weight components from the bulk of IgG. This technique, which can also be run in the reversed sequence, allows phenotyping using complex protein samples from human and other species with high throughput. When urea IEF gels are used, even SDS polyacrylamide gel electrophoresis can be run in the first dimension as shown in Reference 20.

7. Running Conditions

For an IEF run, the film- or glass-supported gel slabs are placed on the cooling plate of the IEF apparatus that has been coated with kerosene for optimal temperature transfer. Water or detergent solutions are not recommended as contact fluids, since they often accumulate some ions by diffusion and become conductive, leading to diversion of the electric field and sparking along the lateral edges of the gels at high voltages.

For non-washed gels and for washed gels with long separation time IEF, as well as for basic gradients and urea gels, filter paper strips soaked in electrode solutions are required between the gel edges and the electrodes. The filter strips provide reservoirs for ionic contaminants in the gels and should be 0.5–1 mm thick and 5 mm wide. Table 1 shows recommended electrode solutions for 0.5 mm thick polyacrylamide gels run with different pH gradients.

Native IEF gels are usually run at 10°C and denaturing IEF gels containing 8 M urea are run at 20°C. Since the isoelectric points of proteins are dependent on temperature, for consistency, it is best to run IEF slab gels at controlled temperature and to provide the temperature information as part of the definition of the isoelectric point.

The pH gradient should be established before the samples are loaded. Therefore, a prefocusing step is necessary to move the carrier ampholytes to their isoelectric points. Samples are applied with small applicator pieces of a cotton/cellulose mixture or with silicone rubber applicator masks or frames following prefocusing. It is highly recommended to evaluate the optimal application point on the pH gradient in an initial trial test.[12] Sample applicator pieces should be removed after the sample entrance phase, but applicator masks or frames can remain on the gel surface during the focusing run. Filter paper is not recommended for sample application, because some proteins irreversibly stick to it and are not released even at high-electric field strengths.

TABLE 1 Electrode Solutions for IEF in 0.5-mm Thin Polyacrylamide Gels

pH gradient	Anode	Cathode
3.5–9.5	0.5 mol/L H_3PO_4	0.5 mol/L NaOH
2.5–4.5	0.5 mol/L H_3PO_4	2% (w/v) ampholytes, pH 5–7
2.5–4.5	0.5 mol/L H_3PO_4	0.4 mol/L HEPES
3.5–5.0	0.5 mol/L H_3PO_4	2% (w/v) ampholytes, pH 6–8
4.0–5.0	0.5 mol/L H_3PO_4	1 mol/L glycine
4.0–6.5	0.5 mol/L acetic acid	0.5 mol/L NaOH
4.5–7.0	0.5 mol/L acetic acid	0.5 mol/L NaOH
5.0–6.5	0.5 mol/L acetic acid	0.5 mol/L NaOH
5.5–7.0	2% (w/v) ampholytes, pH 4–6	0.5 mol/L NaOH
5.0–8.0	0.5 mol/L acetic acid	0.5 mol/L NaOH
6.0–8.5	2% (w/v) ampholytes, pH 4–6	0.5 mol/L NaOH
7.8–10.0	2% (w/v) ampholytes, pH 6–8	1 mol/L NaOH
8.5–11.0	0.2 mol/L histidine	1 mol/L NaOH

When more than 20 μL of sample solution is applied, plastic application frames are very useful. These frames can be prepared by cutting the bottom away from disposable photometer cuvettes. The frames are placed on the gel surface, with the smooth edges down, before prefocusing and remain on the gel throughout the entire separation time. In order to prevent leakage, it is a good idea to dip the smooth edges into a puddle of 100% glycerol before placement on the gel.

During sample entry the voltage must be kept low to prevent aggregation and precipitation of the proteins. When the carrier ampholyte pH gradient is established, the conductivity in the gels becomes rather low. Thus, the electric conditions are best controlled with the voltage settings in the power supply. The closer proteins come to their isoelectric points, the less charged they become. This means, that high electric field strength is required to drive the lightly charged proteins to their isoelectric point and keep them focused there.

8. High-Throughput Analysis

Slab gel IEF is a useful technique for the analysis of large sample numbers. It is easy to apply multiple samples into sample applicator masks on horizontal flatbed gels with microplate-compatible multipipettes. This is, in fact, much easier to perform than sample application on vertical electrophoresis gels. One way to double sample throughput is by positioning a single cathode at the center of the gel slab with two anodes, one on each of the two long edges of the gel parallel to the cathode as shown in Figure 5. For this kind of application, the settings for the power supply have to be modified as compared with the standard settings: to double the current and half the voltage. This is because the configuration establishes two electrical current paths, but halves the

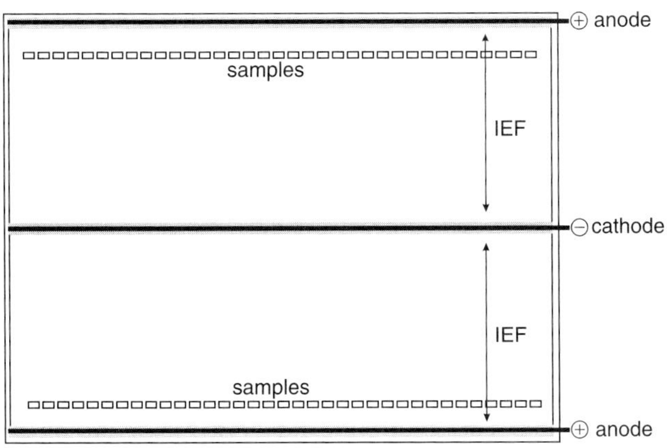

FIGURE 5 Schematic drawing of a setup for high-throughput slab gel IEF.

inter-electrode distance. The settings for power stays the same, because it is the product of voltage and current.

D. Protein Detection

1. Staining

The challenge for staining IEF gels is the fixation of the proteins in the large pore gels, while at the same time efficiently removing the carrier ampholytes. The carrier ampholytes contain primary amino groups and can therefore bind protein stains.

Two ways of staining IEF gels with Coomassie Brilliant Blue are employed: (1) The traditional procedure by fixing the gel with 20% (w/v) trichloroacetic acid and staining with Coomassie Brilliant Blue R-250 dissolved in 40% methanol and 10% acetic acid, with subsequent destaining with 25% methanol and 10% acetic acid. (2) Staining with colloidal Coomassie Brilliant blue G-250 according to Diezel et al.[23] or Blakesley and Boezi.[24] This procedure fixes small proteins and oligopeptides more efficiently than the traditional method, and it is almost odorless.

Fluorescent dyes have the advantage of a wide linear dynamic range, which allows very reliable quantification. The most sensitive fluorescent dye for IEF is SYPRO Ruby[25] (Available from Molecular Probes/Invitrogen, Bio-Rad). However, a fluorescence scanner or a UV table and a camera are needed for visualization.

A number of silver staining methods exist for IEF gels. The most sensitive variant is the ammoniacal silver stain, which is used to detect unconcentrated oligoclonal IgG in cerebrospinal fluid according to Wurster.[45] The protocol for this method can also be found in Reference 12.

2. Zymogram Detection

When enzymes or enzyme inhibitors are separated by IEF under native conditions, they can be detected by active zymogram staining, which is performed by placing the gel into the appropriate substrate solution and coupling the enzymatic reaction with a dye reaction. In most cases, the protocols described for histological studies and electrophoretic separations[26,27] work also for IEF gels. Here it is very useful to use prepolymerized, washed gels and short separation distances, because enzyme activities can suffer by contact with catalysts, non-reacted monomers, and long time exposure to the separation matrix.

3. Immunofixation

When IEF gels of 0.5 mm and less thickness are used, selected protein species can be fixed and detected by immunofixation with specific polyclonal antibodies. This detection method also enhances sensitivity. As an example, two silver stained band patterns of human serum and

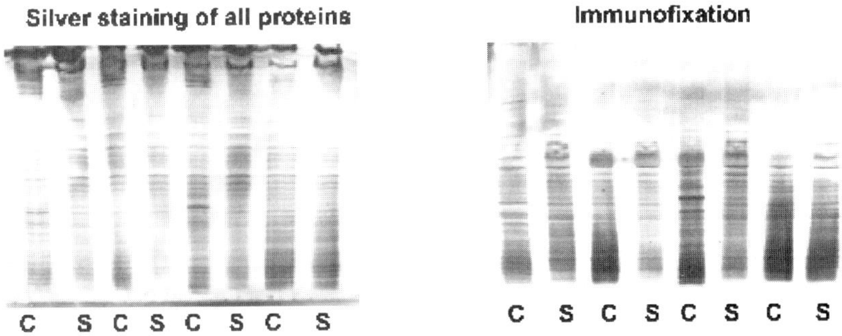

FIGURE 6 IEF of human sera and cerebrospinal fluid in small polyacrylamide gels. Detection of proteins was with silver staining after fixing all proteins with trichloroacetic acid (left) and after fixing only the immunoglobulins by immunofixation with anti-IgG (right). Sample applied on the left gel, 4 μL = 80 ng IgG lane, sample applied on the right gel, 1 μL = 20 ng IgG/lane (IgG concentration was measured with nephelometry).

cerebrospinal fluid separated in small polyacrylamide gels are shown in Figure 6. The same samples were applied to the two gels, but at different volumes: 4 μL of each sample were applied on the trichloroacetic acid fixed gel and 1 μL per sample on the immuno-fixed gel. Immunofixation is mostly employed with agarose gels.

4. Immunoblotting

Electroblotting of proteins from IEF gels is problematic because the soft gels are generally supported on film supports, and the proteins are uncharged at their isoelectric points and must acquire charges to allow for electrophoretic transfer. Towbin et al.[28] described a very efficient procedure for immunoblotting of focusing gels using pressure transfer.

5. Isoelectric Point Markers

For native IEF, the isoelectric points of the focused proteins can be determined with the help of a calibration curve, established by plotting the known isoelectric points of co-focused marker proteins over the gel distance. If marker proteins disturb the focusing patterns or when denaturing conditions are used, non-protein low-molecular weight compounds, such as amphoteric dyes,[29] have been used.

E. Immobilized pH Gradients

Immobilized pH gradients are prepared by co-polymerizing acrylamide derivatives of buffer compounds containing acidic and basic groups with the polyacrylamide network.[30] The gradient-forming buffers become grafted directly into the polyacrylamide gel matrix. Immobilized pH gradient gels are cast in essentially the same manner as used for casting porosity gradient gels using recipes and methods which are published, for instance, in the IPG book by Righetti[31] and in Reference 32 (see also Chapter 4 of this book).

With current chemistries, IEF in immobilized pH gradients is restricted to polyacrylamide gels. Although it is conceivable to create immobilized pH gradients in agarose gels using agarose-buffer derivatives, charged agarose chains would separate from each other under the influence of the electric field and the gel matrix would be destroyed.

The concept of using fixed buffering groups in polyacrylamide slab gels allows creation of extremely flat pH gradients down to 0.02 pH units/cm for very high resolution.[33] Because the buffering groups are fixed and cannot migrate like carrier ampholytes, these gels have a very low conductivity allowing use of very high electric fields. Most importantly, they are not subject to cathode drift and can be run for the extended times at high voltage necessary for focusing many proteins.

Immobilized pH gradient gels are cast on support films and washed with water after polymerization. It is important to remove all mobile ions from the gel matrices before use to prevent their build up at the electrodes with subsequent shortening of the gradients. During the washing process, the gels swell considerably which aids in clearing away mobile ions. Washed gels are then dried down to paper-thin films and rehydrated before use. Rehydration is done either in glass cassettes or in plastic trays with a defined liquid volume. Immobilized pH gradient slab gels are usually 0.5 mm thick; thicker gels are more complicated to wash, dry, and rehydratate.

Because the buffering groups are grafted to the gel, no adverse effects occur when holes are punched into the gel for improved sample application. This measure is advantageous for detergent-containing samples and prevents lateral band spreading for samples with a high molar urea content.[34] Furthermore, there are no lateral effects in IEF when the gel slabs are cut into narrow (3–4 mm wide) strips before rehydration. IPG strips have become the standard medium for denaturing IEF as the first dimension in high-resolution two-dimensional electrophoresis.[35]

V. AGAROSE GELS

Catsimpoolas[36] was first to use agarose slab gels for IEF: he poured agarose gels on standard microscope slides to form 2-mm-thick layers. For sample application, a small pit was made in the middle of the layer. The advantages of agarose are the absence of polymerization catalysts and the large pore size, which allows larger molecules to migrate without restrictions. The agarose polysaccharide forms hydrogen-bonded double helices, which join laterally and create relatively thick filaments, resulting in large pore size gels with good mechanical stability. Thus, it is the ideal matrix for separating large proteins like IgA and IgM.

Electroendosmosis problems prevented the agarose-gel method from becoming widely accepted during the early stages of its development. About 10 years after the first experiments, agaroses with very low electroendosmosis have been available from several different suppliers. One approach to reducing electroendosmotic effects is through the removal of sulfonic acid groups from the agarose chains.[37] A stronger electroendosmosis suppression effect is achieved by compensating of acidic groups with addition of alkaline groups. However, completely electroendosmosis-free agarose does not exist. Therefore, it is important to include 10% (w/v) sorbitol in the gels, which reduces the electroendosmotic water flow considerably by increasing the viscosity of the gel solution.

A. Gel Preparation

An agarose IEF gel is prepared by heating aqueous 10% sorbitol to boiling, and then adding 0.8% (w/v) agarose powder. After the agarose has melted, the mixture is cooled to about 78°C and carrier ampholytes are added. This mixture is pipetted into a glass cassette with a 0.5 mm thick gasket. One glass plate is covered with a special polyester foil (GelBond; available from Amersham Biosiences, Bio-Rad, Cambrex Bio Science), to which the agarose gel binds. After the gel has solidified, the gel-foil layer is removed from the cassette and placed overnight in a humidity chamber in a refrigerator. This treatment is necessary to allow the final desired agarose structure to form.

Agarose–urea gels are not practically possible, because urea disrupts the hydrogen bonding of the filaments, thereby producing a very soft matrix.

B. Running Conditions

For agarose focusing, the electrode solutions are different from those used for polyacrylamide gels (see Table 2). Because of electroendosmotic effects, for agarose gels, the use of sample applicator pieces is not recommended; only sample applicator masks should be employed. During IEF some electroendosmotic effects are usually observed: the anodal electrode strip dries out and needs to be replaced by a new one after 30 min; at the same time the cathodal strip becomes very wet and should be dried with filter paper. Usually the separation is run at 10°C, but cryoproteins, like IgM, would precipitate at low temperatures and are therefore run at 37°C.

TABLE 2 Electrode Solutions (Mol/L) for IEF in Agarose Gels

pH gradient	Anode	Cathode
3.5–9.5	0.25 acetic acid	0.25 NaOH
2.5–4.5	0.25 acetic acid	0.4 HEPES
4.0–6.5	0.25 acetic acid	0.25 NaOH
5.0–8.0	0.04 glutamic acid	0.25 NaOH

C. Protein Detection

1. Staining

After fixation with 20% trichloroacetic acid, the agarose IEF gel is first washed twice with 10% acetic acid and 25% methanol in water and then dried by placing some layers of filter paper on its surface and pressing the gel with a glass plate under 1–2 kg weight for 10 min. After the gel has been completely dried in a heating cabinet, it can be stained similar to polyacrylamide gel. Because agarose gels will not reswell in staining solutions, staining and destaining of dried thin layers works much faster than for thicker gel layers. If silver staining is required, a one-step colloidal staining method for dried gels is employed.[38] Silver staining of agarose gels had already been introduced 8 years before silver staining of polyacrylamide gels.[39] However, the sensitivity of this technique is considerably lower than the method used for polyacrylamide gels.

2. Immunofixation

Agarose gels are the ideal matrices for immunofixation, because the pore sizes are large enough for the antibody molecules to diffuse into the gel layer. Figure 7 shows a separation of neuraminidase treated human sera after immunofixation with polyclonal antibodies against plasminogen isoenzymes and subsequent Coomassie Brilliant Blue staining according to Leifheit et al.[40]

VI. DEXTRAN GELS

Delincee and Radola[41] introduced the granulated dextran gel as an anticonvective medium for IEF. The method is particularly useful for preparative applications[42] in up to 10-mm-thick Sephadex gel beds. Granulated beds of Sephadex are free of catalysts and have almost no retardation effect. Slurries of dextran are poured into metal or plastic frames to form rectangular beds. Electrode contact is made at the ends of the beds and samples are pipetted directly onto the bed. Optimal use of granulated beds is critically dependent on the correct water content of the slurry: if it is too wet, it will not act as an anticonvective medium,

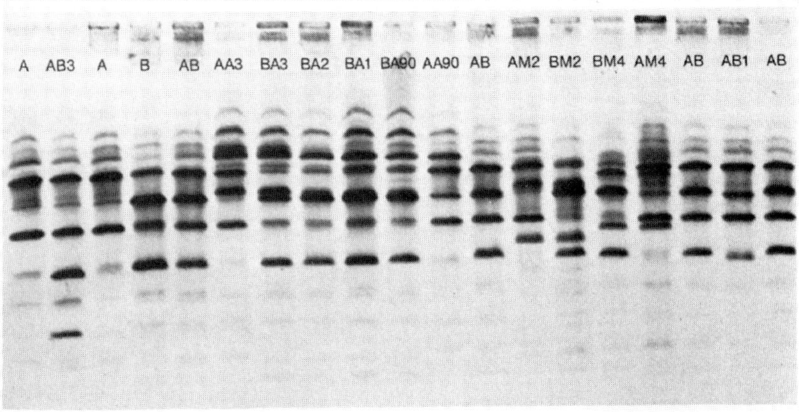

FIGURE 7 IEF of neuraminidase-treated human sera in an agarose gel with a pH gradient 5–8. Coomassie Brilliant Blue staining after immunofixation with polyclonal antibodies against plasminogen isoenzymes (reproduced after Leifheit et al.[40]).

and if it is too dry, the gel bed will crack during the run. After focusing, the proteins are detected by blotting the surface of the bed with a filter paper and staining this print with a protein detection dye such as Coomassie Brilliant Blue. Protein bands identified on the print are first bracketed by segments of a horizontal grid that fits inside the slurry frame. The slurry containing the bands of interest are simply scooped out with a spatula and the proteins are separated from the slurry beads by centrifugation or filtration. Ziegler and Köhler[43] combined the benefits of granulated gels for focusing very large proteins without steric hindrance with the ease of handling polymerized gels. They mixed acrylamide and crosslinker into the Sephadex slurry and polymerized the gel after IEF by spraying TEMED and ammonium persulfate onto the gel. The proteins are visualized by staining, but cannot be further analyzed because they become modified and fixed into the matrix by the polymerization process.

Lately, the technique of IEF in granulated gels has experienced a revival as a very useful method for prefractionation of complex protein mixtures according to the isoelectric points for high-resolution two-dimensional electrophoresis in proteomics.[44]

VII. EXPERIMENTAL PROTOCOLS: POLYACRYLAMIDE SLAB GEL IEF

A. Carrier Ampholyte Polyacrylamide Gel IEF

The standard protocol for slab gel IEF is presented here.

1. Equipment

Multiphor (see Figure 2), Casting Cassette, GelBond PAG, Power Supply capable of delivering 3000 V at >5 mA, Recirculating Chiller, Staining Tray, etc.

2. Gel Casting

(a) Stock Solutions

Carrier Ampholytes: Carrier ampholytes are usually supplied as 40% (w/v) solutions, although some narrow range carrier ampholytes are supplied at 20% (w/v). Some suppliers do not specify the dry weight content of their ampholytes. In the recipe shown in Table 3, it is assumed that the carrier ampholyte stock is at 40% (w/v).

Acrylamide, Bis solution ($T = 40\%$, $C = 3\%$): Dissolve 38.8 g of acrylamide and 1.2 g of *N,N'*-methylenbisacrylamide, in 60–70 mL of distilled, deionized water. Stir until all grains have dissolved, then filter the solution through paper. When stored in a dark place at 4°C (refrigerator), the solution can be kept for 1 week. This solution is commercially available from several suppliers.

Caution! Acrylamide and *N,N'*-methylenbisacrylamide are toxic in the monomer form. Avoid skin contact and do not pipette by mouth.

Ammonium Persulfate Solution 40% *(w/v)*: Dissolve 400 mg of ammonium persulfate in 1 ml of distilled water.This solution is stable for 1 week when stored in the refrigerator (4°C).

TEMED (N,N,N',N'-tetramethylethylenediamine) is not necessarily needed for polymerization. However, since carrier ampholytes from different suppliers can be chemically very different, TEMED should be added in order to guarantee polymerization effectiveness and consistency. TEMED is used neat. It should not be older than 1 year and should be kept in the refrigerator.

(b) Preparation of the Cassette: Pour a few milliliters of distilled water on the blank glass plate and place the support film (GelBond PAG Film®) with the hydrophobic side on the water puddle. Move the film until the short edges are flush with the short edges of the glass plate and one of

TABLE 3 Gel Recipe for a 0.5-mm Thin IEF Slab Gel 5% *T*, 3% C Gel of 25 × 12 cm

Acrylamide Bis solution (40%*T*, 3%C)	1.9 mL
Monoethylenglycol (100%)	1.5 mL
Carrier ampholytes pH 3–10 (40% w/v)	750 µL
TEMED (100%)	8 µL
Distilled water	10.8 mL
Total	15.0 mL

15 µL ammonium persulfate solution (40% w/v) is added immediately before filling of the cassette

the long edges protrudes over a long edge of the glass plate by 1mm. Press the film down on the glass plate with a roller. Place a U-shaped gasket cut from a 0.5-mm-thick silicone rubber on the film and put another glass plate on top of it. The upper glass plate should be treated once with RepelSilane™ to allow easy removal of the plate from the softgel surface after polymerization. When the sandwich is clamped, it forms a cassette as shown in Figure 8. Tilt the cassette vertically for filling.

(c) Polymerization: Prepare fresh monomer solution according to the recipe in Table 3. After thorough mixing of the solution, deaerate it with a vacuum pump for 5 min to remove oxygen. Add 15 μL of ammonium persulfate solution and mix the monomer solution thoroughly, but carefully, without creating bubbles. Immediately pipette the monomer solution into the cassette as shown in Figure 9. A setup like this can also be used for casting agarose gels.

Overlay the upper edge of the gel with a few hundred microliters of distilled water to prevent oxygen diffusion into the upper gel. The solution should be allowed to polymerize overnight before the gel is used. The gel can be wrapped in a plastic foil and stored in a refrigerator for several weeks.

3. Sample Preparation

For Coomassie Brilliant Blue staining, adjust the protein concentration to around 1–3 mg/mL with distilled water. The salt concentration should not exceed 50 mM; apply 10–20 μL to the gel. Also apply 10 μL of pI marker proteins (pH 3–10) to at least two lanes.

4. Isoelectric Focusing

In Figure 3 a schematic setup for slab gel IEF is shown. Set the temperature of the thermostatic circulator to 10°C. Pipette 3 ml of kerosene on the cooling plate. Remove the gel from the cassette and place it on the

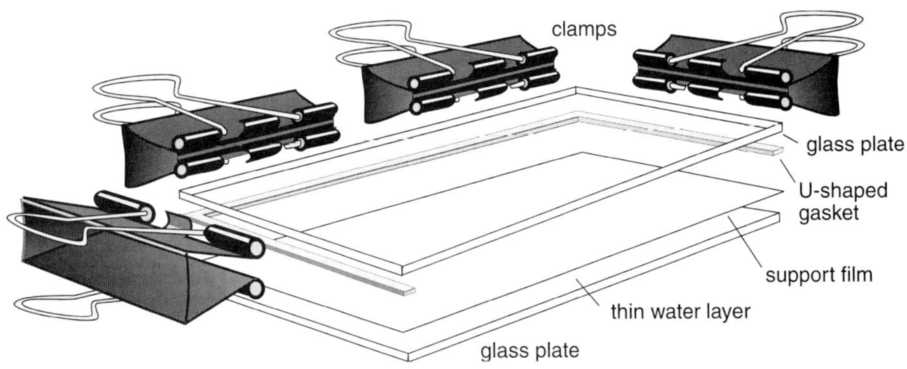

FIGURE 8 Assembly of a cassette for casting an IEF slab gel on a film support. The U-shaped gasket can also be glued to the surface of the upper glass plate.

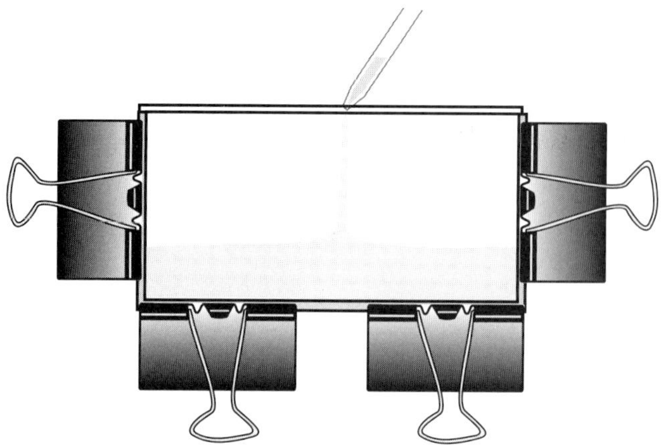

FIGURE 9 Pipetting the monomer solution into the cassette for the polymerization of an IEF slab gel.

TABLE 4 Power Supply Settings for a 0.5-mm Thin IEF Slab Gel 5% *T*, 3% *C* Gel of 25 × 12 cm

	Time (min)	Maximal voltage (V)	Maximal current (mA)	Maximal power (W)
Prefocusing	20	700	20	10
Sample entrance	30	500	20	10
Separation	90	2000	20	10
Band focusing	10	2500	5	15

cooling plate with the gel facing upward. The kerosene should distribute uniformly under the gel's support foil.

(a) Electrode Strips: Soak one electrode strip with 0.5 M phosphoric acid and place it on the anodal edge of the gel layer. Soak a second electrode strip with 0.5 M sodium hydroxide and place it on the cathodal edge. Blot excess liquid from the electrode strips with filter paper before applying them to the gel.

(b) Focusing: The power supply settings are listed in Table 4. Figure 10 shows several ways to apply sample to a flatbed gel. Mode 1 is only recommended for agarose gels; mode 2 shows sample applicator pieces, which can be applied after the prefocusing phase; for modes 3 and 4 the silicon rubber application masks or glass rings should be applied before prefocusing to avoid leaking of the sample, because the formation of ridges on the gel surface can already start in this phase. The choice of the

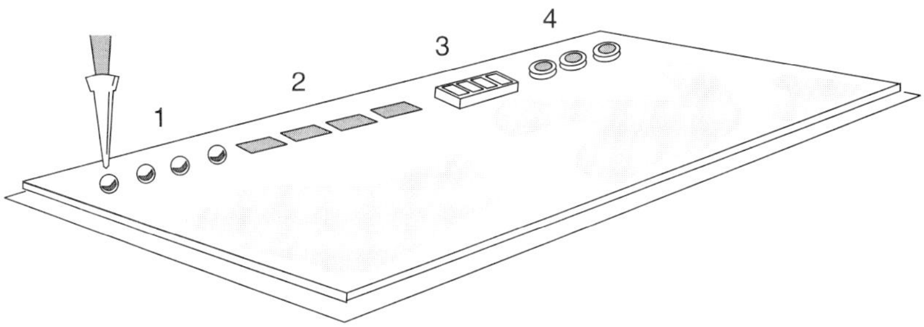

FIGURE 10 **Different means of sample application on a flat-bed slab gel for IEF. From Reference 32. (1) Direct application as a droplet, (2) applicator pieces, (3) silicon rubber application mask, (4) glass rings.**

optimal sample application mode is dependent on the sample and should be tested for each new sample type.

The position of the optimal sample application point is dependent on the kind of sample and should be selected with the help of a reference in the literature or laboratory manual, or must be determined with a step trial test (see above). In general, do not apply samples where they are expected to focus. Remove the sample applicator pieces after the sample entry phase; the masks or glass rings should remain on the gel surface. Run the phases for separation and band sharpening consecutively without interruption.

5. Protein Detection by Colloidal Coomassie Brilliant Blue Staining

The following colloidal Coomassie Brilliant Blue staining procedure is the most useful for IEF slab gels, as explained above: Dissolve 2 g of Coomassie Brilliant Blue G-250 in 1 L of distilled water Add 1 L of 1 M sulfuric acid (1 M; 55.5 mL of concentrated H_2SO_4 per liter) while stirring. After further stirring for 3 h, filter the solution through paper, and then add 220 mL of 10 M sodium hydroxide (10 M; 88 g NaOH in 220 mL) to the brown filtrate. Finally, add 310 mL of 100% (w/v) trichloroacetic acid and mix well. The solution will turn green. Colloidal sols of Coomassie Brilliant Blue G-250 are commercially available for protein staining.

Fixing and staining is performed in one step: 3 h at 50°C or overnight at room temperature in the colloidal sol. Later the acid is washed out by soaking the gel in water for 1–2 h. The green bands become blue and more intense as the water drives the dye molecules into the proteins.

B. Immobilized pH Gradient IEF

The equipment is the same as for carrier ampholyte IEF gels.

1. Equipment

Multiphor (see Figure 2), Casting Cassette, Gradient Maker, GelBond PAG, Power Supply capable of delivering >3000 V with the minimum current safety switch turned off, Recirculating Chiller, Staining Tray,

2. Gel Casting

Preparation of immobilized pH gradient gels is much more complicated and prone to errors than making laboratory-cast carrier-ampholyte IEF gels. Only a small selection of ready-made IPG gel slabs is commercially available (from Amersham Biosciences only). Therefore, the entire procedure is described here.

(a) Stock Solutions

Immobiline® II 0.2 molar stock solutions:

Acids: p*K*-3.6, 4.6.

Bases: p*K*-6.2, 7.0, 8.5, and 9.3.

The solutions are stabilized against autopolymerization and hydrolysis and have a shelflife of at least 12 months when stored in the refrigerator (4–8°C). Immobilines® II should not be frozen!

Acrylamide, Bis solution ($T = 40\%$, $C = 3\%$): Dissolve 38.8 g of acrylamide and 1.2 g of *N,N'*-methylenbisacrylamide, in 60–70 mL of distilled, deionized water. Stir until all grains have dissolved, then filter the solution through paper. When stored in a dark place at 4°C (refrigerator) the solution can be kept for 1 week. This solution is commercially available from several suppliers.

Caution! Acrylamide and *N,N'*-methylenbisacrylamide are toxic in the monomer form. Avoid skin contact and do not pipette by mouth.

Ammonium persulfate solution 40% (*w/v*): Dissolve 400 mg of ammonium persulfate in 1 mL of distilled water.

This solution is stable for one week when stored in the refrigerator (4°C).

TEMED (N,N,N',N'-tetramethylethylenediamine) (100%): TEMED should not be older than 1 year and should be kept in the refrigerator.

4 M HCl: Dissolve 33.0 mL of concentrated HCl in 67.0 mL of distilled water.

(b) Preparation of the Cassette: Pour a few milliliters of distilled water on the blank glass plate and place the support film (GelBond PAG Film®) with the hydrophobic side on the water puddle. Move the film until the short edges are flush with the short edges of the glass plate and one of the long edges protrudes over a long edge of the glass plate by 1mm. Press the film down on the glass plate with a roller. Place a U-shaped gasket cut from a 0.5-mm-thick silicone rubber on the film and put another glass plate on top of it. The upper glass plate should be treated once with RepelSilane™ to allow easy removal of the plate from the soft gel surface after polymerization. When the sandwich is clamped, it forms a cassette as shown in Figure 8. Chill the cassette in a refrigerator in order to delay the

start of polymerization. This measure is taken to ensure that the density gradient has settled before polymerization begins. Note that settling of a gradient in a 0.5-mm-thin cassette is much slower than for a 1mm cassette.

(c) Casting a pH Gradient Gel and Polymerization: Immobilized pH gradients are cast in a similar way as porosity or additive gradient gels. A large number of Immobiline recipes are found in References 31 and 32. Prepare two fresh monomer solutions according to Table 5. The pH gradient is stabilized by a glycerol density gradient.

The gradient maker consists of two communicating chambers (Figure 11). First the light, basic, solution is pipetted into the rear cylinder, the channel between the cylinders is opened very briefly and immediately closed again to fill up the channel with light solution, thus avoiding an air bubble barrier between the two solutions. The dense, acidic solution and a stirrer bar are placed in the front cylinder, the mixing chamber. A compensation bar is placed into the rear cylinder, the reservoir, to balance the volume of the magnetic stirrer and the difference in density: 25% glycerol is added to the dense solution and 5% to the light one so that it is easier to overlay the gel solution in the cassette with water before polymerization.

The casting cassette is removed from the refrigerator and placed close to the gradient maker, with the glass plate holding the film facing

TABLE 5 Recipe for Starting Solutions for one 0.5-mm Thick IPG Gel (pH 4 to 10), $T = 4\%$, $C = 3\%$ of 25 × 12 cm

Solutions	pH 4.0 (Dense)	pH 10.0 (Light)
Immobiline p*K* 3.6	551 μL	—
Immobiline p*K* 4.6	—	57 μL
Immobiline p*K* 6.2	227 μL	25 μL
Immobiline p*K* 7.0	45 μL	244 μL
Immobiline p*K* 8.5	167 μL	79 μL
Immobiline p*K* 9.3	—	179 μL
Acrylamide Bis solution (40% *T*, 3% *C*)	750 μL	750 μL
Glycerol (87 %)	2.2 mL	400 μL
TEMED (100 %)	4 μL	4 μL
Fill up with distilled water mix carefully and measure the pH with pH paper	to 7.5 mL	to 7.5 mL
With 4 mol/L HCl, titrate to pH 7		10 μL*
With TEMED, titrate to pH 7		8 μL*
Total	7.5 mL	7.5 mL
Ammonium persulfate solution (40% w/v) is added in the gradient maker immediately before filling the cassette	8 μL	8 μL

*Experimental values.

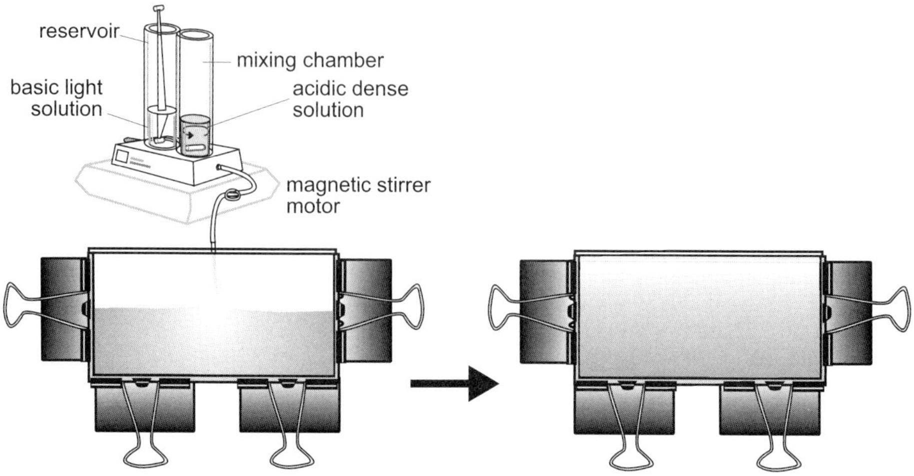

FIGURE 11 **Casting a linear pH gradient gel with a gradient maker. The gradient settles after about 10 min.**

the operator. This allows a better control of the tip of the tubing, which is inserted into the center of the cassette between the upper edges. At this time, the ammonium persulfate is added and mixed first with the dense solution by briefly turning on the magnetic stirring motor, and second with the light solution using the compensation bar. The magnetic stirrer motor is turned on and adjusted to a speed producing a small vortex is obtained. Fast rotation must be avoided in order to prevent the development of air bubbles. The channel between the chambers is opened and the clamp at the front tubing is released. The gel solution will flow into the cassette through the tubing under the influence of gravity.

Overlay the upper edge of the gel with a few hundred microliters of distilled water to prevent oxygen diffusion into the upper gel. Do not use alcohol-containing overlay solutions. The gel is allowed to polymerize for 2 h at room temperature. At first the gradient will not be straight. It takes about 10 min for the gradient to settle completely (see Figure 11).

(d) Gel Washing, Drying, and Rehydratation: Remove the gel from the cassette and wash it four times for 15 min, each in 0.5 L of distilled water, on a shaker. In order to avoid curling of the drying gel, incubate it for another 15 min in 1.5% (v/v) glycerol in distilled water. Dry the gel at room temperature in a dust-free cabinet. When the gel is dry, immediately cover it with an inert plastic film and store it in a plastic bag in a freezer.

Rehydratation can either be performed in a vertical cassette or in a reswelling tray as shown in Figure 12. The vertical rehydration cassette also allows rehydration with an urea gradient perpendicular to the pH gradient as described in Reference 19. The gel casting cassette can also

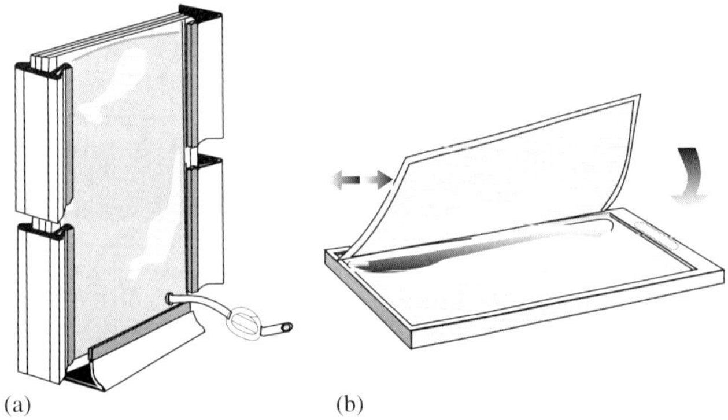

FIGURE 12 **Methods for rehydratation of an IPG gel: (a) rehydration cassette; (b) reswelling tray.**

be used for rehydratation. When the gel is just rehydrated in distilled water, the matrix is completely reconstituted.

3. Sample Preparation

For Coomassie Brilliant Blue staining, adjust the protein concentration to around 1–3 mg/mL with distilled water. The salt concentration should not exceed 50 mM. Apply 10–20 μL to the gel. Apply 10 μL of pI marker proteins (pH 3–10) to at least two lanes.

4. Isoelectric Focusing

In Figure 3 a schematic setup for slab gel IEF is shown. Set the temperature of the thermostatic circulator to 10°C. Pipette 3 mL of kerosene on the cooling plate. Place the gel on the cooling plate with the gel facing upward, and with the acidic side at the anode. The kerosene should distribute uniformly under the gel's support foil. Usually no electrode strips are needed.

(a) Sample Application: For focusing in IPG gels, the sample must be applied without prefocusing in order to use the initial current to transport the sample proteins into the gel. All sample application modes shown in Figure 10 can be used. Additionally, in IPG gel IEF, holes can be punched into the gel because the gradient is fixed. Also, in IPG gels, the position of the optimal sample application point is dependent on the kind of sample and should be selected with the help of a reference in the literature or laboratory manual, or must be determined with a step trial test.

(b) Focusing: For IPG gels, 10°C is the optimal temperature. When 8M urea has been added to the rehydration solution, 20°C is chosen. Because the pH gradient is already established in the gel, the power supply settings are very simple: one phase with the maximum set to 3500 V, 1.0 mA and 5.0 W. The minimum separation time is dependent on the pH

61. Daban, J. R., Bartolome, S. and Samso, M. Use of the hydrophobic probe Nile red for the fluorescent staining of protein bands in sodium dodecyl sulfate-polyacrylamide gels. *Anal. Biochem.* 199:169–174, 1991.
62. Malone, J., Radabaugh, M. R., Leimgruber, R. M. and Gerstenecker, G. S. Practical aspects of fluorescent staining for proteomic applications. *Electrophoresis* 22:919–932, 2001.
63. Lauber, W. M., Carroll, J. A., Dufield, D. R., Kiesel, J. R., Radabaugh, M. R. and Malone, J. P. Mass spectrometry compatibility of two-dimensional gel protein stains. *Electrophoresis* 22:906–918, 2001.
64. Berggren, K., Chernokalskaya, E., Steinberg, T. H., Kemper, C., Lopez, M. F., Diwu, Z., Haugland, R. P. and Patton, W. F. Background-free, high sensitivity staining of proteins in one- and two-dimensional sodium dodecyl sulfate-polyacrylamide gels using a luminescent ruthenium complex. *Electrophoresis* 21:2509–2521, 2000.
65. Berggren, K. N., Schulenberg, B., Lopez, M. F., Steinberg, T. H., Bogdanova, A., Smejkal, G., Wang, A. and Patton, W. F. An improved formulation of SYPRO Ruby protein gel stain: comparison with the original formulation and with a ruthenium II tris (bathophenanthroline disulfonate) formulation. *Proteomics* 2:486–498, 2002.
66. Steinberg, T. H., Agnew, B. J., Gee, K. R., Leung, W. Y., Goodman, T., Schulenberg, B., Hendrickson, J., Beechem, J. M., Haugland, R. P. and Patton, W. F. Global quantitative phosphoprotein analysis using Multiplexed Proteomics technology. *Proteomics* 3:1128–1144, 2003.
67. Steinberg, T. H., Pretty On Top, K., Berggren, K. N., Kemper, C., Jones, L., Diwu, Z., Haugland, R. P. and Patton, W. F. Rapid and simple single nanogram detection of glycoproteins in polyacrylamide gels and on electroblots. *Proteomics* 1:841–855, 2001.
68. Anderson, N. L., Nance, S. L., Tollaksen, S. L., Giere, F. A. and Anderson, N. G. Quantitative reproducibility of measurements from Coomassie Blue-stained two-dimensional gels: analysis of mouse liver protein patterns and a comparison of BALB/c and C57 strains. *Electrophoresis* 6:592–599, 1985.
69. Giometti, C. S., Gemmell, M. A., Nance, S. L., Tollaksen, S. L. and Taylor, J. Detection of heritable mutations as quantitative changes in protein expression. *J. Biol. Chem.* 262:12764–12767, 1987.
70. Giometti, C. S., Gemmell, A. M., Tollaksen, S. L. and Taylor, J. Quantitation of human leukocyte proteins after silver staining: a study with two-dimensional electrophoresis. *Electrophoresis* 12:536–543, 1991.
71. Blomberg, A., Blomberg, L., Norbeck, J., Fey, S. J., Larsen, P. M., Larsen, M., Roepstorff, P., Degand, H., Boutry, M., Posch, A. and Gorg, A. Interlaboratory reproducibility of yeast protein patterns analyzed by immobilized pH gradient two-dimensional gel electrophoresis. *Electrophoresis* 16:1935–1945, 1995.
72. Lopez, M. F. and Patton, W. F. Reproducibility of polypeptide spot positions in two-dimensional gels run using carrier ampholytes in the isoelectric focusing dimension. *Electrophoresis* 18:338–343, 1997.
73. Molloy, M. P., Brzezinski, E. E., Hang, J., McDowell, M. T. and VanBogelen, R. A. Overcoming technical variation and biological variation in quantitative proteomics. *Proteomics* 3:1912–1919, 2003.
74. Motulsky, H. *Intuitive Biostatistics*, Oxford University Press, NY, pp. 195–204, 1995.
75. Lachin, J. Introduction to sample size determination and power analysis for clinical trials. *Contro. Clin. Trials* 2:93–113, 1981.
76. Petricoin, E. F., Ardekani, A. M., Hitt, B. A., Levine, P. J., Fusaro, V. A., Steinberg, S. M., Mills, G. B., Simone, C., Fishman, D. A., Kohn, E. C. and Liotta, L. A. Use of proteomic patterns in serum to identify ovarian cancer. *Lancet,* 359:572–577, 2002.
77. Liotta, L. A., Ferrari, M. and Petricoin, E. Clinical proteomics: written in blood. *Nature* 425:905, 2003.
78. Pieper, R., Gatlin, C. L., Makusky, A. J., Russo, P. S., Schatz, C. R., Miller, S. S., Su, Q., McGrath, A. M., Estock, M. A., Parmar, P. P., Zhao, M., Huang, S. T., Zhou, J.,

Wang, F., Esquer-Blasco, R., Anderson, N. L., Taylor, J. and Steiner, S. The human serum proteome: display of nearly 3700 chromatographically separated protein spots on two-dimensional electrophoresis gels and identification of 325 distinct proteins. *Proteomics* 3:1345–1364, 2003.

79. Taylor, S. W., Fahy, E. and Ghosh, S. S. Global organellar proteomics. *Trends Biotechnol.* 21:82–88, 2003.
80. Jung, E., Heller, M., Sanchez, J.-C. and Hochstrasser, D. F. Proteomics meets cell biology: the establishment of subcellular proteomes. *Electrophoresis* 21:3369–3377, 2000.
81. http://us.expasy.org/ch2d/2d-index.html.
82. Gygi, S. P., Corthals, G. L., Zhang, Y., Rochon, Y. and Aebersold, R. Evaluation of two-dimensional gel electrophoresis-based proteome analysis technology. *Proc. Natl. Acad. Sci. USA* 97:9390–9395, 2000.
83. Westbrook, J. A., Yan, J. X., Wait, R., Welson, S. Y. and Dunn, M. J. Zooming-in on the proteome: very narrow-range immobilised pH gradients reveal more protein species and isoforms. *Electrophoresis* 22:2865–2871, 2001.
84. Wilkins, M. R., Gasteiger, E., Sanchez, J.-C., Bairoch, A. and Hochstrasser, D. F. Two-dimensional gel electrophoresis for proteome projects: the effects of protein hydrophobicity and copy number. *Electrophoresis* 19:1501–1505, 1998.
85. Molloy, M. P. Two-dimensional electrophoresis of membrane proteins using immobilized pH gradients. *Anal. Biochem.* 280:1–10, 2000.
86. Santoni, V., Molloy, M. and Rabilloud, T. Membrane proteins and proteomics: un amour impossible? *Electrophoresis* 21:1054–1070, 2000.
87. Anderson, N. L. and Anderson, N. G. The Human Plasma Proteome. *Mol. Cell. Proteomics* 1:845–867, 2002.
88. Molloy, M. P., Herbert, B. R., Walsh, B. J., Tyler, M. I., Sanchez, J.-C., Hochstrasser, D. F., Williams, K. L. and Gooley, A. A. Extraction of membrane proteins for two-dimensional electrophoresis by differential solubility. *Electrophoresis* 19:837–844, 1998.
89. Pedersen, S. K., Harry, J. L., Sebastian, L., Baker, J., Traini, M. D., McCarthy, J. T., Manoharan, A., Wilkins, M. R., Gooley, A. A., Righetti, P. G., Packer, N. H., Williams, K. L. and Herbert, B. R. Unseen proteome: mining below the tip of the iceberg to find low abundance and membrane proteins. *J. Proteome Res.* 2:303–311, 2003.

7

SOME PRACTICES AND PITFALLS OF SAMPLE PREPARATION FOR ISOELECTRIC FOCUSING IN PROTEOMICS

BEN HERBERT

Proteome Systems Ltd., 1/35-41 Waterloo Road, North Ryde, Sydney, NSW 2113, Australia

I. INTRODUCTION

Classical isoelectric focusing (IEF) was mainly concerned with separations of soluble proteins, and at that time it was satisfactory to load proteins dissolved in water or dilute solutions of carrier ampholytes on to the IEF columns or gels. Sufficient solubilization of insoluble proteins was in many cases achieved by dissolving proteins in solutions containing detergents and (or) chaotropes.[1] Initially, IEF was used to study intact proteins, but two related considerations led to the adaptation of completely denatured proteins for IEF separations. The first of these considerations was a need to reduce the ambiguities in separation patterns

Handbook of Isoelectric Focusing and Proteomics
D. Garfin and S. Ahuja, editors.

brought about by the realization that conformational isomers of the same protein can attain different pI values. The second, and perhaps more fundamental, consideration derives from the need to relate protein sequence information to genomic data. Thus, reducing agents were incorporated in sample preparation techniques in order to reduce the disulfide bridges of cystine residues to cysteine sulfhydryls.

The advent of the proteomic approach to protein biochemistry with its strong reliance on two-dimensional polyacrylamide gel electrophoresis (2-DE) and mass spectrometry, brought with it a re-examination of sample preparation methods for IEF. 2-DE places exacting demands on sample preparation for IEF because it is exquisitely sensitive to artifacts that can change the charges on proteins. In addition, mass spectrometry can detect very small differences in polypeptide masses. It is therefore incumbent on sample preparation methods that they do not introduce artifacts in order that charge and mass differences among polypeptides can be attributed to post-translational modifications and not to sample handling. Sample preparation for IEF can be considered as a subset of sample preparation for 2-DE. The vast majority of improvements in sample preparation for IEF has arisen as methods designed for 2-DE. Consequently, the discussion in this chapter is centered on methods that were devised for and tested by 2-DE. In particular, the physical methods that have been developed for pre-fractionating complex protein samples prior to 2-DE are not discussed, rather some findings are presented regarding the advantages and disadvantages of various detergents, chaotropes, and reducing agents that are used to treat protein samples immediately prior to IEF.

The greatest challenge in proteomics is to identify reagents, or combinations of reagents, that will solubilize the broadest range of proteins and maintain their solubility during the entire 2-DE procedure. Standard methods for IEF rely on non-ionic or zwitterionic reagents to disrupt protein complexes and denature proteins to their constituent polypeptide monomers. A common sample preparation solution for 2-DE consists of 8 M urea, 4% CHAPS, 50 mM DTT, 0.2% carrier ampholytes, and 0.001% Bromophenol Blue. (CHAPS is a zwitterionic detergent,[2] DTT is dithiothreitol, a sulfhydryl reducing agent (percentages are w/v), and Bromophenol Blue is for tracking the electrophoresis.) With many types of samples, spot quantity and quality in 2-DE is improved by replacing Bromophenol Blue with Coomassie Brilliant Blue R-250[3] (0.001% made from a 1% stock in 10% isopropyl alcohol), especially at the basic end of the pH gradient. Because IEF separates proteins based on isoelectric point, the single most powerful solubilizing reagent, sodium dodecyl sulfate (SDS), is not normally usable unless it can be displaced by an used IEF-compatible detergent prior to the focusing run. If SDS is used to extract proteins from cells, tissues, or fluids, it must be diluted at least 8–10-fold in an IEF solution containing, for example CHAPS, or it can cause anomalous focusing and horizontal streaking. To approximate the denaturing power of boiling SDS under reducing conditions, IEF practitioners have relied on various cocktails of chaotropes, surfactants, and reducing agents.

Chaotropes, urea being the most common for IEF, disrupt the hydrogen bonding between water and a protein's surface to cause partial unfolding which exposes the (hydrophobic) interior of the protein. This partial unfolding often compromises the protein's solubility in aqueous solution, thus the requirement for detergents, which are often called surfactants. It is normal to have at least one surfactant present in the IEF cocktail to help in solubilizing the hydrophobic domains that are exposed as a result of denaturation.

Even small amounts (20 mM) of ionic substances prolong the electrophoresis and are not generally compatible with steady-state IEF. Thus, the use of strong ionic detergents such as SDS is not recommended and IEF is restricted in use to non-ionic or zwitterionic detergents. Traditionally, non-ionic surfactants such as Triton X-100 and octyl glucoside have been used, however, more recently, these have been superseded by the sulfobetaine class of surfactants such as CHAPS, an amido-sulfobetaine.[1,2]

Finally, the standard IEF sample cocktail includes reducing agents which break disulfide bonds to enable complete protein unfolding and denaturation. The two main types of reducing agents used are the free-thiol reagents such as mercaptoethanol and dithiothreitol and the phosphines, a group of trivalent phosphorous compounds. The traditional free-thiol compounds are used at high concentrations (20–100 mM) and work by displacing the equilibrium toward the breakage of disulfides. However, the reagents are charged at alkaline pH and reducing conditions are almost impossible to maintain during IEF.[4] The non-charged phosphines, such as tributyl phosphine (TBP), provide improved reducing conditions and thus improved focusing for some samples.[4] However, even the phosphines fail to provide reducing conditions for the overnight run times required for equilibrium focusing in immobilized pH gradients (IPGs). It is becoming apparent that the ultimate sample preparation method for disulfides is to alkylate the reduced cysteines prior to IEF, which has the added advantage of avoiding reducing conditions during and after the IEF run. The chemistry of cysteine has a number of pitfalls which will be discussed in detail in this chapter.

Since 1996 a number of publications have reported and reviewed the use of novel reagents such as thiourea and new sulfobetaine surfactants, which improve protein solubilization prior to IEF.[1,4–7] Thiourea at 2 M, in combination with urea at 7 M, produces a far more chaotropic sample solution than the conventional 8 M urea. However, this increased chaotropic power required a new class of surfactants to cope with the highly denaturing environment. Rabilloud developed several new surfactants, the best of which are named amido-sulfobetaine 14 (ASB-14) and

C7bZ0.[5,6] In combination with urea and thiourea, either of these detergents provides a formidable level of solubilizing power. The increased solubility obtainable with these new reagents coupled with the high resolution of separation on narrow-range IPGs has significantly increased the total number of resolvable proteins in 2-DE analysis. However, the increased number of proteins solubilized from a single sample causes difficulties in attempts to separate them on a single 2-D gel and complexity reduction via prefractionation is essential.

As proteomics matures there is a renewed awareness of co- and post-translational modifications, as these are clearly the mechanisms that generate protein complexity given the relatively small number of genes in the human genome. Thus, in separation sciences the pressure is on to ensure that artifactual protein modifications are eliminated or at least minimized.

II. REDUCTION AND ALKYLATION

In 2-DE, the standard procedure adopted up to the present calls for reduction prior to the IEF/IPG step, followed by a second reduction/alkylation step in the equilibration solution between the first and second dimension, in preparation for the SDS–PAGE step. This protocol is far from being optimal, due to incomplete reduction during the IEF and often results in a large number of spurious multimeric spots, due to "scrambled" disulfide bridges between like and unlike chains. Due to the negative charge on the –SH group of typical reducing agents such as DTT, this compound can act as a buffer and it will migrate inside the pH gradient (toward the anode) until it is arrested, by protonation, at around pH 7. Thus, artifacts arising from incomplete reduction are more often observed in the alkaline portion of the IPG. Even tributyl phosphine, a strong non-thiol reducing agent, does not appear to have the reducing power to maintain all proteins as monomeric polypeptides during the IEF. This might possibly be because TBP is so volatile that it evaporates during extended IEF runs. The situation is even worse in conventional carrier ampholyte IEF, where a steady-state distribution of thiol reducing agent can shorten a pH 3–10 gradient to a pH 3–7.5 span. We have shown in a recent series of papers that the number of this type of artifactual spots can be impressively large even in the case of simple polypeptides such as the human α- and β-globin chains, which possess only one (α-) or two (β-) –SH groups, respectively.[8–10] Figure 1 compares human α- and β-globin chains solubilized in 7 M urea, 2 M thiourea, reduced with TBP (left panel) and alkylated with acrylamide (right panel). The dry IPG strips (pH 6–11, 6.5 cm long and 4 mm wide, home-made) were rehydrated with the globin sample (ca. 4 mg/ml, 150 μl) for 4 h (passive sample loading).

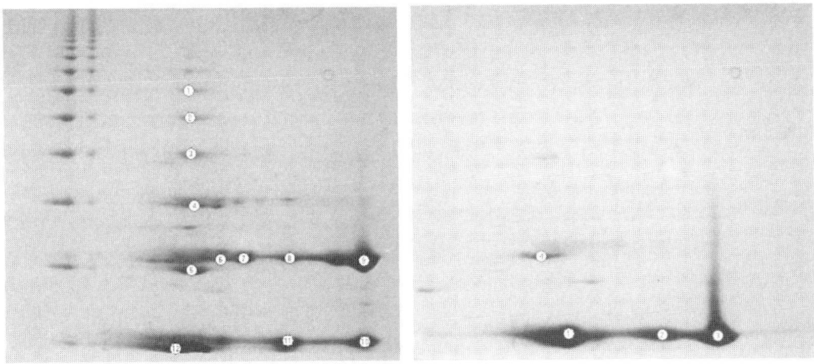

FIGURE 1 Effect of alkylation prior to IEF. Purified human α- and β-globin chains were reduced with tributyl phosphine (left) or reduced with tributyl phosphine and alkylated with acrylamide (right) prior to the IEF step of 2-DE. The white spots show where proteins were excised for MS analysis. The strings of spots on the left panel (reduced but not alkylated) are attributed to multimers of the two globin chains. The 2-DE gel positions of monomeric globin chains are shown on the right panel (reduced and alkylated). The high-M_r spot on the right image is a carbonic anhydrase contaminant.

The numbered white spots in the 2-D images in Figure 1 are the spot excision marks where proteins have been cut for MS analysis. On the right 2-D gel, the higher M_r spot is a carbonic anhydrase contaminant.

In addition, failure to alkylate proteins prior to the IEF step can result in a substantial loss of spots on the 2-D gel, probably due to the fact that proteins, at their pI values, regenerate disulfide bridges with concomitant formation of aggregates which become entangled with and trapped within the polyacrylamide gel fibers. This strongly inhibits their transfer to the subsequent SDS–PAGE gel (data not shown). Even the addition of large quantities of reducing agents and subsequent alkylation in the IPG equilibration step, in the conventional protocol, is ineffective because SDS strongly inhibits –SH alkylation.[9] Similar results, supporting the use of alkylation, have recently been published.[11] In this work, Rabilloud and co-workers found that cysteine blocking highly increased resolution and decreased streaking, especially in the basic region of their 2-D gels.

Poor alkylation efficiency can be obtained using iodoacetamide as the alkylating agent, especially in the presence of thiourea, which acts as a scavenger of iodoacetamide. If iodoacetamide is dissolved in a solubilizing mixture containing thiourea, but in the absence of sample proteins, it will be destroyed quite rapidly by thiourea.[10] When thiourea is used in the solubilizing solution, iodoacetamide should be added as a powder or in water just prior to alkylation. Prolonged alkylation reactions (~24 h) with iodoacetamide should be avoided because they can give rise to modifications of lysine and other amino acids such as methionine.[9,10,12]

To simplify the methodology of reduction and alkylation a very appealing option is to use acrylamide (or an activated double bond) as an alkylating agent instead of iodoacetamide.[8–10] In fact, at pH values at which alkylation is customarily performed, iodoacetamide and acrylamide have similar reactivity rates on ionized –SH groups. Acrylamide does not react with TBP, so the reduction and alkylation can be done in a single step (containing 5 mM TBP and 15 mM acrylamide), resulting in a significant saving in time. Furthermore, acrylamide does not seem to react with thiourea or amino acid groups other than cysteine.

III. BETA ELIMINATION OF CYSTEINE

Apart from preventing disulfide reformation, alkylation has another benefit in preventing artifactual β-elimination of cysteine during electrophoresis. β-elimination, which results in the loss of an H_2S group (34 Da) from Cys residues of proteins focusing in the alkaline pH region, has recently been reported.[13] The confirmation of β-elimination occurence was obtained by trapping an alkaline protein in an electric field using a multicompartment electrolyzer (MCE; Proteome Systems Ltd., Sydney). A sample of lysozyme was solubilized in 8 M urea and focused in a four-chamber MCE for up to 48 h. The chambers of the MCE were separated by acrylamide/Immobiline membranes (7.5%*T*, 10%*C*) cast onto glass fiber discs (2 mm × 24 mm, pore size 2.7 μm). The immobilized pH discs had pH values of 3.0, 8.0, and 11.0 constructed according to the instructions provided by the manufacturer. Lysozyme was loaded into the alkaline pH 8.0–11.0 chamber of the MCE. Focusing was for 4 h at 100 V followed by 44 h at constant 1 W and temperature was monitored regularly using a thermocouple thermometer. A control sample under static, non-electric field conditions was solubilized in 8 M urea in sodium borate (pH 9.0) and maintained at an equivalent temperature to the MCE sample for 48 h. Conductivity measurements of the MCE chamber solutions and the control sample were regularly made using a micro-conductivity meter. Figures 2a and b show a series of mass spectra over a 24 h time course of MCE electrophoresis.

At time zero (Figure 2a) the starting unalkylated lysozyme shows the correct mass of 14313 Da. After 6 h in the electric field, the spectrum reveals two additional compounds, one centered at *m/z* 14278 (corresponding to the loss of 34 Da) and one at *m/z* 14215 (loss of 98 Da). Such M_r decrements are consistent with the loss of one and three H_2S groups, respectively. In Figure 2b, the expanded mass spectra show the starting lysozyme in the upper panel (the peak at 7159 is the doubly charged ion) and the massive degradation after 24 h electrolysis in the lower panel. The process taking place is β-elimination from Cys residues, transforming

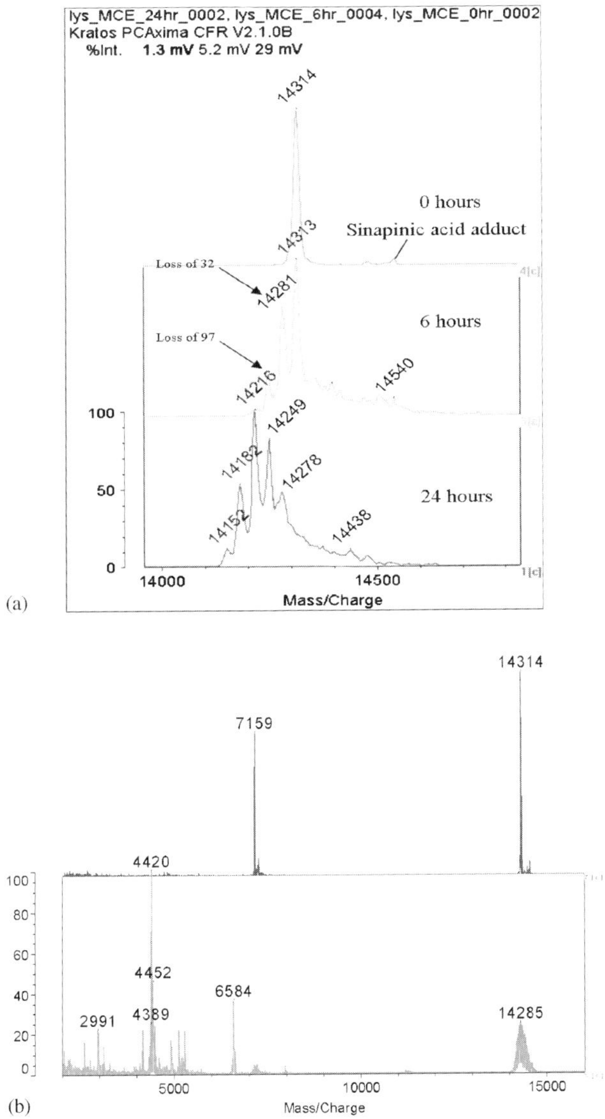

FIGURE 2 **Electrophoretic β-elimination of cysteine. MALDI-TOF-MS spectra of lysozyme after 0 h (upper tracing), 6 h (middle spectrum), or 24 h (lower tracing) of electrophoresis in an MCE are shown. After 24 h of electrophoresis, nearly all of the native lysozyme has been replaced by a series of degradation products consistent with the loss of up to five H_2S units.**

them into dehydro-alanine residues. If the process is continued for 24 h, the peak of the intact protein disappears, giving rise to a heterogeneous spectrum of peaks exhibiting progressive mass losses down to *m/z* 14152. This seems to be an electrically driven process, since the control lysozyme solution, which was left standing in a test tube for the same time intervals at the same pH, did not show any degradation (not shown). Upon such an elimination event, a dehydro-alanine residue is generated at the Cys site. In turn, the presence of a double bond in this position elicits lysis of the peptide bond, often generating a number of peptides of fairly large size from an intact protein. The first process seems to be favored by the electric field, probably due to the continuous harvesting of the SH^- anion produced.

The only remedy found for this degradation pathway is the reduction and alkylation of all Cys residues prior to their exposure to the electric field. Alkylation appears to substantially reduce both β-elimination and the subsequent amido bond lysis. Figure 3 shows a time course of mass spectra of alkylated lysozyme which has been electrolysed in an MCE under the same conditions as the unalkylated lysozyme in Figure 2.

It can be appreciated that alkylation is strongly protective against β-elimination. In the upper tracing (control) only the peak of intact, octa-alkylated lysozyme appears at *m/z* 14882 (the peak at *m/z* 7425 being the doubly charged ion). After 24 h in the electric field, essentially the same two peaks are visible (lower spectra), with only traces of degraded products at *m/z* 14773 and 2712.

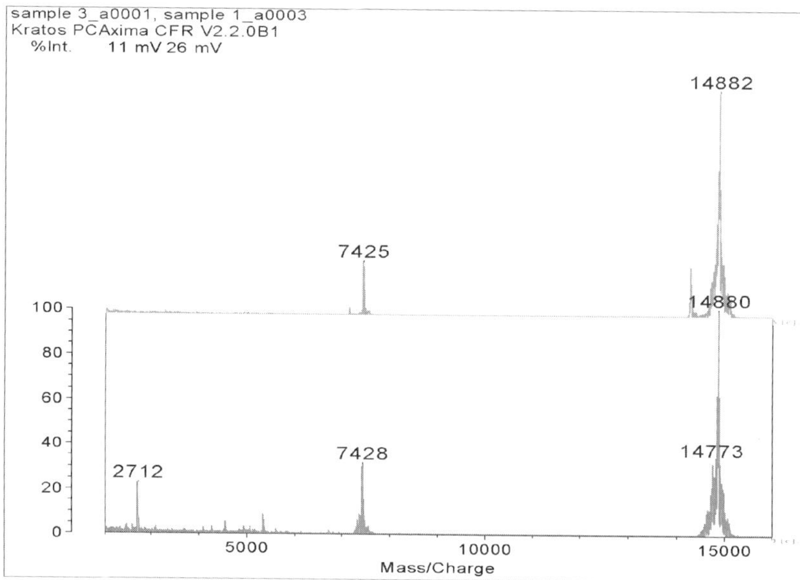

FIGURE 3 **Alkylation prevents electrophoretic β-elimination of cysteine. MALDI-TOF-MS spectra of octa-alkylated lysozyme after 0 h (upper spectrum) or 24 h (lower spectrum) of electrophoresis in an MCE. The peak at *m/z* 7425 is that of the doubly charged ion. Alkylation strongly quenches the degradation of the protein into peptides.**

IV. CARBAMYLATION

As discussed above, urea is the most commonly used chaotropic agent. Thiourea is increasingly being used in combination with urea in order to exploit its high denaturing ability. In solution, urea is in equilibrium with ammonium cyanate[14] as shown in Figure 4.

Urea solutions that are initially cyanate-free can be prepared by using mixed-bed ion exchangers, but the concentration of ammonium cyanate slowly increases over time until equilibrium is again reached. However, if a cyanate scavenger such as the ε-amino group of lysine is present, the formation of cyanate will continue unabated until the scavenger is completely consumed. At temperatures below 37°C, the degradation of urea proceeds slowly and concentrations of cyanate do not reach problematic levels within the time of most sample preparation procedures. Temperatures above 37°C accelerate the rate at which ammonium cyanate is produced and thus should be avoided when preparing protein samples in urea.

Cyanate has been shown to react with nucleophilic groups such as the amino terminus of the protein, the amino side chains of lysine and arginine residues, and the sulfhydryl groups of cysteine residues.[15,16] The reaction occurs more rapidly under alkaline conditions when the nucleophilic groups are deprotonated and thus more reactive. The relative reactivity of the residues is dependent on their individual pK_a values. The free base forms of aliphatic amines, such as the ε-amino group of lysine, are present at very low concentrations below pH 8. The carbamylation reaction of amines by isocyanic acid is strongly pH-dependent and a pH of 8.5–9.5 is usually optimal for modifying lysine residues. In contrast, the R-amino groups at amino termini of the protein are neutral, with pK_a values of ~7, and may be selectively modifed by reaction at near neutral pH[17]. The carbamylation modification results in an increase of 43 AMU relative to the unmodified protein or peptide.[18]

One potential drawback of the reduction and alkylation procedure described above is the fact that the reaction requires alkaline pH to proceed. Unfortunately, the alkaline pH also ensures that lysine residues are deprotonated and reactive toward isocyanate. Under static conditions, i.e., not in an electric field, the cyanate is free to react with lysine and

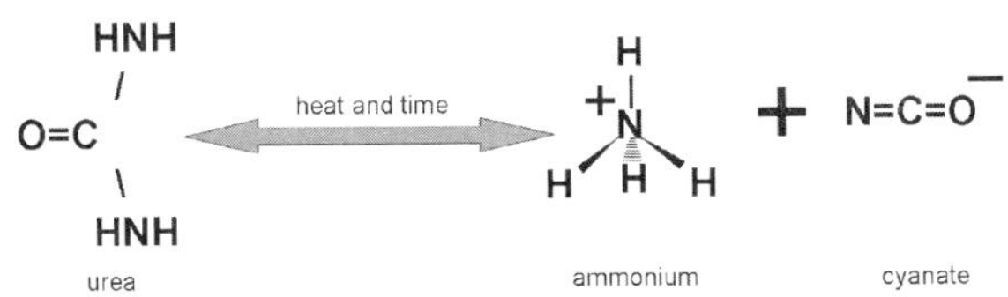

FIGURE 4 **Degradation pathway of urea. The breakdown of urea to ammonium cyanate is driven by heat, time, and pH.**

cause artifactual modifications. The basic recommendation for sample preparation is to minimize extraction time and ensure that the sample is kept as cool as possible during extraction and subsequent storage without causing urea precipitation and at least below 37°C at all times. Storage of urea-containing extracts should always be at 4°C for short periods and frozen for long-term storage. Despite the chemical potential for carbamylation under static conditions, it is clearly shown in Reference 19 that the progression of the reaction is quite slow and carbamylation was not detectable within the first 12 h at room temperature.

Under dynamic conditions in an electric field during IEF, the kinetics of carbamylation is very different compared with the situation described above. During electrophoresis, the charged products of urea degradation are rapidly transported to the electrodes, thus affording them minimal opportunity to react with amino groups on proteins and peptides. The protective effect of the electric field is shown in Figure 5, where the spectra show MS analysis of a myoglobin peptide trapped in an alkaline chamber of a multicompartment electrolyser.

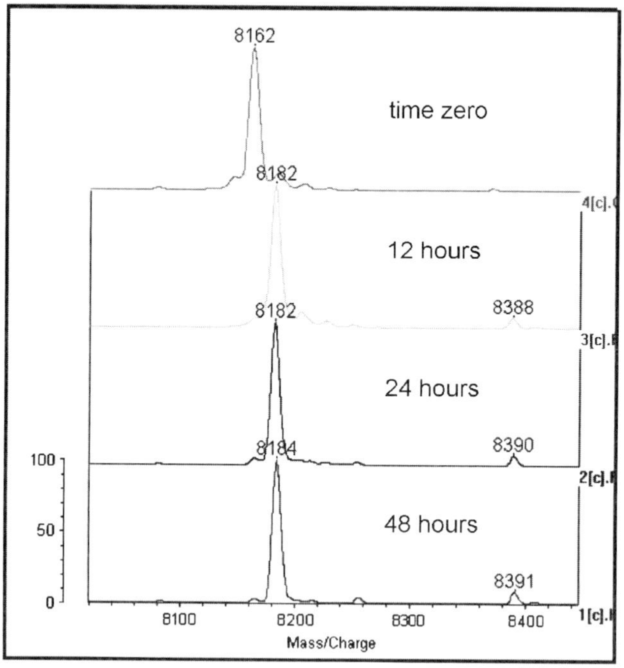

FIGURE 5 Extent of carbamylation of a myoglobin peptide in the presence of 8 M urea in an electric field. The analysis was by MALDI-TOF-MS. Times of electrophoresis in an MCE are 0, 12, 24, and 48 h. There is no carbamylation of the sample kept in an electric field. The mass increase from 8162 to 8182 is due to the conversion of terminal methionine into homoserine lactone.

The mass increase from 8162 to 8182 is due to conversion of the terminal methionine into homoserine lactone. No peaks corresponding to the addition of 43 AMU were observed in the myoglobin peptide trapped in the MCE, even after 48 h of electrolysis in the alkaline chamber. The peptide was the only buffering species present in the alkaline MCE chamber. The pH of the chamber was close to 9.3, defined by the calculated pI of the peptide. The additional masses observed at 8390–8391 are unidentified contaminants.

V. STABLE ISOTOPE LABELING-BASED QUANTITATION

A number of groups have adopted the isotopic labeling advantages of the isotope-coded affinity tag (ICAT) method and applied it to 2-D gels. Smolka et al.[20] used the ICAT reagents to enable quantitative protein profiling of a yeast differential display. The method is based on the same strategy as the LC-MS ICAT experiments, that two or more isotopically encoded samples can be separated concurrently in the same gel. This works because proteins labeled with isotopically different affinity tag reagents precisely co-migrate during two-dimensional electrophoresis. In a variant of this approach, two groups have reported the use of deuterated acrylamide and normal acrylamide in isotopically coded 2-D gel experiments.[21,22] These tagging methods employ labels that couple with the reactive sulfhydryl groups of cysteine residues. So, the alkylation chemistry of cysteine turns out to be at the center of proteomics, from preventing artifactual spots and protein autodigestion to isotopically encoded quantitation.

VI. SAMPLE HOMOGENIZATION AND NUCLEIC ACID REMOVAL

Nucleic acids are polyanions and bind to many proteins via electrostatic interactions. They, along with lipids, some polysaccharides, and ionic contaminants in protein samples give rise to severe streaking and other deleterious effects in IEF. The ideal sample preparation for 2-D gels combines as many steps as possible designed to produce the best quality separations while minimizing sample losses. The need to remove contaminants governs the types of sample preparation methods that can be used with tissue or cellular samples.

A. Mechanical Methods

Although each sample type poses its own challenges for effective preparation for IEF, for the most part, standard methods of protein extraction are effective. Common artifacts are easily avoided by means of a few

contained up to seven transmembrane domains, they were not hydrophobic as judged by GRAVY values and were readily solubilized after the more abundant soluble proteins had been removed. Recent papers have used a more aggressive type of sequential extraction where the membrane is stripped of peripheral proteins by incubation in sodium carbonate at pH 11. In a study of E. Coli, 80% of the integral outer-membrane proteins were detected on a single gel using the sodium carbonate, high pH membrane stripping method.[35] A study of yeast, made by combining literature-confirmed and predicted membrane proteins, identified a total of 105 integral membrane proteins, which was 33% of the 323 unique proteins identified.[31]

REFERENCES

1. Rabilloud, T. Solubilization of proteins for electrophoretic analysis. *Electrophoresis* 17:813–829, 1996.
2. Hjelmland, L. H. and Chrambach, A. Electrophoresis and electrofocusing in detergent containing media: a discussion of basic concepts. *Electrophoresis* 2:1–11, 1981.
3. Vilain, S., Cosett, P., Charlionet, R., Hubert, M., Lange, C., Junter, G.-A. and Jouenne, T. Substituting Coomassie Brilliant Blue for bromophenol blue in two-dimensional electrophoresis buffers improves the resolution of focusing patterns. *Electrophoresis* 22:4368–4374, 2001.
4. Herbert, B. R., Molloy, M. P., Gooley, A. A., Walsh, B. J., Bryson, W. G. and Williams, K. L. Improved protein solubility in two-dimensional electrophoresis using tributyl phosphine as reducing agent. *Electrophoresis* 19:845–851, 1998.
5. Chevallet, M., Santoni, V., Poinas, A., Rouquie, D., Fuchs. A., Keiffer, S., Rossignol, M., Lunardi, J. and Rabilloud, T. New zwitterionic detergents improve the analysis of membrane proteins by two-dimensional electrophoresis. *Electrophoresis* 19: 1901–1909, 1998.
6. Rabilloud, T., Blisnick, T., Heller, M., Luche, S., Aebersold, R., Lunardi, J. and Braun-Breton, C. Analysis of membrane proteins by two-dimensional electrophoresis: comparison of the proteins extracted from normal or Plasmodium falciparum-infected erythrocyte ghosts. *Electrophoresis* 20:3603–3610, 1999.
7. Santoni, V., Molloy, M. and Rabilloud, T. Membrane proteins and proteomics: un amour impossible? *Electrophoresis* 21:1054–1070, 2000.
8. Herbert, B., Galvani, M., Hamdan, M., Oliveri, E., McCarthy, J., Pedersen, S. and Righetti, P. G. Reduction and alkylation of proteins in preparation of two-dimensional map analysis: why, when and how? *Electrophoresis* 22:2046–2057, 2001.
9. Galvani, M., Hamdan, M., Herbert, B. and Righetti, P. G. Alkylation kinetics of proteins in preparation for two-dimensional maps: a matrix assisted laser desorption/ionization-time of flight-mass spectrometry investigation. *Electrophoresis* 22:2058–2065, 2001.
10. Galvani, M., Rovatti, L., Hamdan, M., Herbert, B. and Righetti, P. G. Protein alkylation in presence/absence of thiourea in proteome analysis: a matrix assisted laser desorption/ionization-time of flight- mass spectrometry investigation. *Electrophoresis* 22:2066–2074, 2001.
11. Luche, S., Diemer, H., Tastet, C., Chevallet, M., Van Dorsselaer, A., Leize-Wagner, E. and Rabilloud, T. About thiol derivatization and resolution of basic proteins in two-dimensional electrophoresis. *Proteomics* 4:551–561, 2004.

12. Sickmann, A., Dormeyer, W., Wortelkamp, S., Woitalla, D., Kuhn, W. and Meyer, H. E. Indentification of proteins from human cerebrospinal fluid, separated by two dimensional polyacrylamide gel electrophoresis. *Electrophoresis* 21:2721–2728, 2000.
13. Herber, B., Hopwood, F., Oxley, D., McCarthy, J., Laver, M., Grinyer, J., Goodall, A., Williams, K., Castagna, A. and Righetti, P. G. Beta-elimination: an unexpected artefact in proteome analysis. *Proteomics* 3:826–831, 2003.
14. Shapiro, R. Prebiotic cytosine synthesis: a critical analysis and implications for the origin of life. *Proc. Natl. Acad. Sci. USA* 96:4396–4401, 1999.
15. Stark, G. R. Reactions of cyanate with functional groups of proteins. IV. Inertness of aliphatic hydroxyl groups. Formation of carbamyl- and acylhydantoins. *Biochemistry* 4:2363–2367, 1965.
16. Stark, G. R. Reactions of cyanate with functional groups of proteins. III. Reactions with amino and carboxyl groups. *Biochemistry* 4:1030–1036, 1965.
17. Anderson, N. L. and Hickman, B. J. Analytical techniques for cell fractions. XXIV. Isoelectric point standards for two-dimensional electrophoresis. *Anal. Biochem.* 93:312–320, 1979.
18. http://www.ionsource.com/Card/carbam/carbam.htm
19. McCarthy, J., Hopwood, F., Oxley, D., Laver, M., Castagna, A., Righetti, P. G., Williams, K. and Herbert, B. Carbamylation of proteins in 2-D electrophoresis—myth or reality? *J. Proteome Res.* 2:239–242, 2003.
20. Smolka, M., Zhou, H. and Aebersold, R. Quantitative protein profiling using two-dimensional gel electrophoresis, isotope-coded affinity tag labeling, and mass spectrometry. *Mol. Cell Proteomics* 1:19–29, 2002.
21. Gehanne, S., Cecconi, D., Carboni, L., Righetti, P. G., Domenici, E. and Hamdan, M. Quantitative analysis of two-dimensional gel-separated proteins using isotopically marked alkylating agents and matrix-assisted laser desorption/ionization mass spectrometry. *Rapid Commun. Mass Spectrom.* 16:1692–1698, 2002.
22. Sechi, S. A method to identify and simultaneously determine the relative quantities of proteins isolated by gel electrophoresis. *Rapid Commun. Mass Spectrom.* 16:1416–1424, 2002.
23. Castellanos-Serra, L. and Paz-Lago, D. Inhibition of unwanted proteolysis during sample preparation: evaluation of its efficiency in challenge experiments. *Electrophoresis* 23:1745–1753, 2002.
24. Link, A. J. (Ed.) 2-D *Proteome Analysis Protocols*. Humana Press, Totowa, NJ, 1999.
25. Palacino, J. J., Sagi, D., Goldberg, M. S., Krauss, S., Motz, C., Klose, J. and Shen, J. Mitochondrial dysfunction and oxidative damage in Parkin-deficient mice. *J. Biol. Chem.* 279:18614–18622, 2004.
26. Razin, S. and Rozansky, R. Mechanism of the antibacterial action of spermine. *Arch. Biochem. Biophys.* 81:36–54, 1959.
27. Hoopes, B. C. and McClure, W. R. Studies on the selectivity of DNA precipitation by spermine. *Nucleic Acids Res.* 9:5493–5504, 1981.
28. Pohl, T. Concentration of proteins and removal of solutes. In *Methods in Enzymology*, Vol. 182 (M. P. Deutscher, Ed.) Academic Press, San Diego, pp. 68–83, 1990.
29. Mastro, R. and Hall, M. Protein delipidation and precipitation by tri-*n*-butlyphosphate, acetone, and methanol. *Anal. Biochem.* 273:313–315, 1999.
30. Mewes, H. W., Frishman, D., Gruber. C., Geier, B., Haase, D., Kaps, A., Lemcke, K., Mannhaupt, G., Pfeiffer, F., Schüller, C., Stocker, S. and Weill, B. MIPS: a database for genomes and protein sequences. *Nucl. Acids Res.* 28:37–40, 2000.
31. Pedersen, S. K., Harry, J. L., Sebastian, L., Baker, J., Traini, M. D., McCarthy, J. T., Manoharan, A., Wilkins, M. R., Gooley, A. A,, Righetti, P. G., Packer, N. H., Williams, K. L. and Herbert, B. R. Unseen proteome: mining below the tip of the iceberg to find low abundance and membrane proteins. *J. Proteome Res.* 2:303–311, 2003.

32. Wilkins, M. R., Gasteiger, E., Sanchez, J. C., Bairoch, A. and Hochstrasser, D. F. Two-dimensional gel electrophoresis for proteome projects: effects of protein hydrophobicity and copy number. *Electrophoresis* 19:1501–1505, 1998.
33. Rabilloud, T. Personal communication.
34. Molloy, M. P., Herbert, B. R., Walsh, B. J., Tyler, M. I., Traini, M., Sanchez, J. C., Hochstrasser, D. F., Williams, K. L. and Gooley, A. A. Extraction of membrane proteins by differential solubilization for separation using two-dimensional gel electrophoresis. *Electrophoresis* 19:837–844, 1998.
35. Molloy, M. P., Herbert, B. R., Slade, M. B., Rabilloud, T., Nouwens, A. S., Williams, K. L. and Gooley, A. A. Proteomic analysis of the Escherichia coli outer membrane. *Eur. J. Biochem.* 267:2871–2881, 2000.

8 PROTEIN DETECTION AND IMAGING IN IEF GELS

WAYNE F. PATTON

Biochemistry Department, Perkin-Elmer Life Sciences, Boston, MA, USA

I. INTRODUCTION

The very earliest attempts at isoelectric fractionation may be attributed to the studies of Ikeda and Suzuki in 1912, whereas modern day carrier ampholyte-mediated isoelectric focusing (IEF) coalesced in the 1950s and 1960s, mainly from Kolin's concept of focusing ions in a continuous pH gradient, Svensson-Rilbe's theoretical construct that proposed an approach to developing stable pH gradients upon applying an electric field, and Vesterberg's actual chemical synthesis of the necessary polydisperse mixture of charged molecules possessing good conducting and buffering capabilities.[1] Due to its relative simplicity in implementation, carrier ampholyte-mediated IEF remains a popular separation approach, being commonly employed in basic research, clinical chemistry, agriculture science, the food industry, and forensics. However, conventional IEF using carrier ampholytes does have several inherent weaknesses, such as the necessity for very low ionic strength operating conditions, a tendency for uneven conductivity and buffering capacity

Handbook of Isoelectric Focusing and Proteomics
D. Garfin and S. Ahuja, editors.

silver staining took its place as the pre-eminent protein detection technique in biological research laboratories worldwide.[14–19] Silver staining allowed, for the first time, non-radioactive detection of proteins in the nanogram range instead of the microgram range.

Two categories of silver staining have found widespread utility for the detection of proteins in polyacrylamide gels, the alkaline silver diamine, and the acidic silver nitrate methods. Both approaches depend upon an oxidation step followed by a reduction step that converts silver ions into metallic silver. The alkaline silver diamine methods originated from histological procedures, using ammonium hydroxide to form soluble silver diamine complexes followed by visualization through reduction of free silver ions with formaldehyde in an acidified developer. On the other hand, the acidic silver nitrate methods found their origins from photographic procedures and depend upon gel impregnation with silver ions at acidic pH, followed by reduction of silver ions to elemental metallic silver at alkaline pH using formaldehyde.

Silver staining procedures are relatively complicated, resulting in numerous solution changes and carefully timed steps.[5,7] Certainly, this is largely responsible for the continued reliance on CBB staining by many researchers. Due to the inherent complexity of silver staining procedures, spot intensities may vary significantly from run to run. The linear dynamic range of silver stain is exceedingly poor, only covering a 10-fold range of protein concentration, making detection of changes in protein expression levels difficult to determine. Finally, standard silver staining procedures require glutaraldehyde and formaldehyde, which alkylate α- and ε-amino groups of proteins. Using silver staining procedures, the inherent advantage of higher detection sensitivity compared with CBB staining is offset by inferior sequence coverage in peptide mass fingerprinting experiments. Mass spectrometry-based analysis can be successfully performed if glutaraldehyde is omitted from the staining procedure, but such modified procedures are plagued by decreased staining sensitivity and uniformity as well as increased gel background. The additional step of destaining silver stained gel bands prior to enzymatic digestion reduces background interference and suppression of signals for MALDI–TOF–MS-based peptide mass analysis.[20,21] Although no perfect solution is yet in hand, research efforts to solve silver stain's incompatibility problems with MS continue.[22]

With respect to IEF in particular, it should be noted that the acidic silver nitrate staining procedures tend to stain basic proteins with slightly lower sensitivity and acidic proteins with higher sensitivity than alkaline silver diamine methods. Proteins separated using immobilized pH gradient gels are very poorly stained using the standard alkaline silver diamine methods. However, high background staining by this method can be minimized through extensive fixing and washing. It can also be noted that due to the volatile nature of ammonia, alkaline silver diamine methods are generally acknowledged as being more susceptible to run-to-run variation than the acidic silver nitrate methods. Silver stain kits are commercially available from the standard electrophoresis supply houses.

IV. REVERSE STAINING

A number of reverse stain methods for the detection of proteins in SDS-polyacrylamide gels have been developed, three of which having gained some measure of popularity among practitioners of protein separations.[5–7] These are the potassium chloride, copper chloride, and zinc chloride reverse stain methods. The latter two stains are available commercially. Reverse stains produce a semi-opaque background on the gel surface, which allows proteins to be detected as transparent zones when gels are viewed on a black background or with proper back illumination. Attractive features of the stains include their short staining protocols (5–15 min to completion) and their ability to preserve the biological activity of the proteins.[23,24] Proteins may also be eluted from gels quite readily by chelation of the metal ions with ethylenediaminetetraacetic acid (EDTA). Reverse stains are thus well suited for detection of proteins, their passive elution from gels and their subsequent analysis by MS. It is noteworthy that among the common non-fluorescent detection methods, zinc-imidazole reverse staining in particular appears to have the fewest drawbacks with respect to use in protein mass profiling experiments. Sequence coverage is generally equivalent to or better than those obtained after CBB staining and since gels are not fixed, peptide yields are also superior.[23,24] However, in terms of quantitative capabilities, the stain is inferior to most other techniques, with the linear dynamic range for protein quantitation restricted to the microgram range.

V. FLUORESCENCE STAINING

With the birth of proteomics in the mid-1990s there came renewed interest in protein detection technologies.[5] The interest was motivated by the need to combine high detection sensitivity with broad quantitation capabilities as well as to provide detection approaches that were more compatible with protein identification techniques, especially MS. Fluorescence-based detection came to the forefront because the detection of a fluorescent signal provides a linear response with respect to the amount of protein over a much wider range than is found for the non-fluorescent alternatives like CBB and silver staining. Among the non-fluorescent detection technologies available, only radiolabeling provides comparable capabilities.

Three categories of fluorescent detection methods have gained prominence in the field of proteomics in recent years. The first category of stains interact with proteins indirectly and non-covalently.[25–29] These fluorophores are virtually non-fluorescent in aqueous solution but become fluorescent upon binding to SDS–protein complexes. Since SDS binds to protein with a fairly constant stoichiometry, protein quantitation by this approach should be more reliable than methods based upon interacting with protein primary amines alone. Prominent stains belonging to this class are Nile Red dye, SYPRO Orange dye, SYPRO Red dye, SYPRO Tangerine dye, hydrophobic fluorescein dyes (5-dodecanoyl amino fluorescein, 5-hexadecanoyl amino fluorescein, and 5-octadecanoyl amino fluorescein), and Deep Purple™ dye.[25–29] Typically, this family of dyes provides detection sensitivities that are equivalent to colloidal CBB staining and rapid silver staining methods. Deep Purple dye and the hydrophobic fluorescein dyes, however, appear to be as sensitive as the highest quality silver staining methods.[25,28]

Surprisingly, fluorescent detection of proteins in IEF gels can be achieved using Nile Red dye as well as with SYPRO Red, Orange and Tangerine dyes.[26,27,29] However, in order to accomplish this, gels must be preincubated in SDS since all of these lipophilic dyes bind to proteins indirectly through this anionic detergent. The principal disadvantages of using these fluorescent dyes to stain IEF gels are that detection sensitivity is often poorer than standard CBB staining and the incubation step in SDS is likely to lead to some loss of protein.

The second category of fluorescent total protein stains comprises the colloidal luminescent transition metal complexes, such as the ruthenium-based SYPRO Ruby dye and the closely related (but not identical) fluorophore, ruthenium II tris (bathophenanthroline disulfonate).[30–36] These stains bind to proteins by a mechanism that is quite similar to CBB staining and are as sensitive as the best silver staining procedures available. The dyes are superior to silver staining with respect to linear dynamic range and downstream compatibility with MS-based protein identification techniques. SYPRO Ruby protein gel stain is notable in that it allows sensitive fluorescence detection of proteins in both IEF and IPG gels.[36] Protein bands are selectively stained while the polyacrylamide gel matrix remains unstained, in an analogous manner as with colloidal CBB staining (see Figure 1). One discrete zone of Ampholine brand carrier ampholytes (Amersham Life Sciences) has been observed to stain strongly with SYPRO Ruby dye, and this artifactual staining has been observed with other stains as well.[36,37] Other carrier ampholytes do not produce a similar staining artifact.

In a quantitative study of 11 isoelectric point marker proteins, it was determined that SYPRO Ruby dye is typically 3–30 times more sensitive than acidic silver nitrate staining and colloidal CBB staining in polyacrylamide IEF gels.[36] Effective staining of agarose IEF gels is also feasible using the fluorescent stain.[36] Agarose may be stained by an identical procedure as used for polyacrylamide gels. An alternative

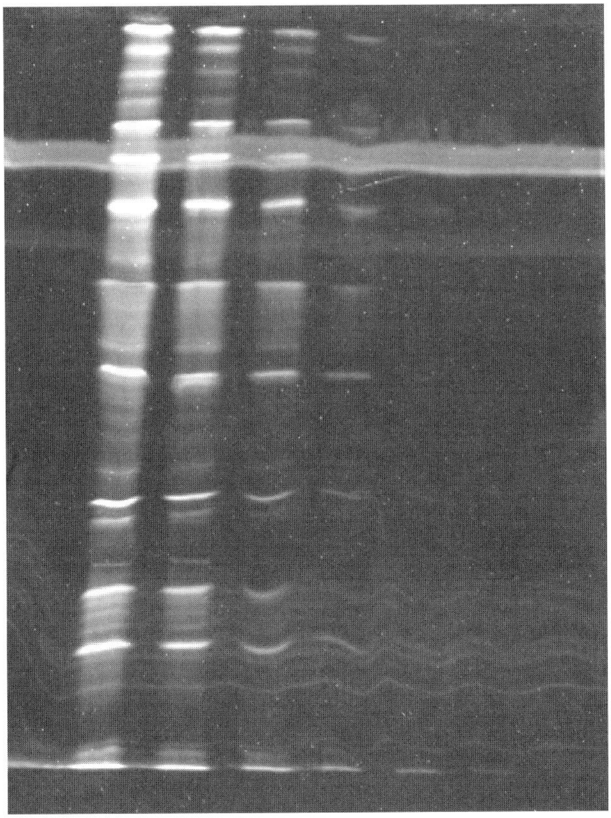

FIGURE 1 **Staining of broad range isoelectric point marker proteins with SYPRO Ruby dye. Electrophoresis was performed on Ampholine PAGplate (pH 3.5–9.5) gels (Amersham Biosciences, Piscataway, NJ). Proteins were loaded in a 3-fold dilution series, with lane 1 containing ~3–7 µg per band. Proteins are, from top to bottom, trypsinogen (pI = .30), lentil lectin (3 bands, 8.65, 8.45, 8.15), myoglobin, basic (7.35), myoglobin, acidic (6.85), human carbonic anhydrase (6.55), bovine carbonic anhydrase (5.85), β-lactoglobulin A (5.20), soybean trypsin inhibitor (4.55), amyloglucosidase (3.50). Courtesy of Dr. Thomas H. Steinberg, Molecular Probes, Inc., Eugene, Oregon.**

method of staining proteins in agarose gels by drying the gels and floating them face down on the surface of the stain solution for 30 min is also appropriate and detection sensitivity using either staining protocol is similar.[36]

The final category of fluorophores commonly used for protein detection in gel electrophoresis is the amine-reactive and sulfhydryl-reactive fluorophores, particularly the cyanine dyes.[38–40] In the most commonly implemented form of difference gel electrophoresis (DIGE), NHS esters of cyanine dyes are employed to pre-label two or three different protein samples prior to running them on the same 2-D gel, allowing the samples to be processed under identical electrophoretic conditions in a type of differential display format.[38] The cyanine dyes have been carefully designed to be charge and mass matched so that they migrate to the same position on a 2-D gel. Although no specific references concerning the use of cyanine dyes in IEF gels alone are available, the labeling method is used with 2-DE routinely and thus, should be considered fully compatible with IEF gel electrophoresis as well.

VI. LABEL-LESS DETECTION

One of the more innovative label-less detection approaches involving IEF gels relies upon laser desorption of proteins directly from dried carrier ampholyte IEF or IPG gels by scanning the laser beam of a MALDI–TOF–MS across the surface of the strip. Sub-picomolar detection sensitivities are achievable with this technique. The approach is referred to as "virtual" 2-D gels, since MS is substituted for SDS–PAGE to construct 2-D protein profiles on a computer screen,[41–43] i.e., the analytical data are displayed as a computer-generated image that is similar to a classical 2-D gel in appearance. A number of methodological artifacts associated with the technique have prevented its widespread adoption, including higher molecular masses than predicted due to difficulty in desorbing proteins from the gel matrix, horizontal streaks arising from variations in baseline slope, artifacts caused by the presence of protein multimers and matrix adducts on the proteins that produce duplicate or triplicate spots in the image, and difficulty in quantifying the amounts of the proteins due to ion suppression phenomenon. The procedure is also currently quite slow, requiring a day to run the gel, 2 days to dry it down and another 2 days to acquire the spectra.

Recently, sensitive direct detection of proteins in polyacrylamide gels has been accomplished by imaging the weak fluorescent signal generated by tyrosine and tryptophan residues in proteins upon illumination with ~280 nm ultraviolet radiation.[44,45] Detection sensitivities of 5 ng protein have been reported using either ultraviolet light laser excitation with a photomultiplier tube-detector or Hg (Xe) lamp excitation with a CCD camera detector.[44,45] Imaging proteins at 230 nm through the peptide bond itself is compromised by the absorption of the polyacrylamide matrix and carrier ampholytes at this wavelength.

VII. POST-TRANSLATIONAL MODIFICATION DETECTION

One of the more important challenges facing the field of proteomics is to reveal rapidly and comprehensively protein post-translational modifications.[7] Two important protein post-translational modifications, glycosylation and phosphorylation, are now readily detectable after PAGE using commercially available kits and instrumentation.

A sensitive green-fluorescent glycoprotein-specific stain, Pro-Q® Emerald 300 dye, detects glycoproteins directly in polyacrylamide gels or on PVDF membranes.[46–48] The dye is conjugated to glycoproteins by the standard periodic acid Schiff's base (PAS) conjugation mechanism using ambient reaction conditions. As little as 300 pg of α_1-acid glycoprotein (40% carbohydrate) can be detected in gels with the dye, and the linear dynamic range of detection extends over a 2–3 order of magnitude range.[46] However, the dye requires UV illumination, rendering it unsuitable for laser-based gel scanners. A related stain, Pro-Q Emerald 488 dye, allows detection of glycoproteins using visible light excitation sources, but unfortunately the alternate stain is about 10-fold less sensitive than the UV-excitable dye.[47]

Pro-Q Diamond phosphoprotein stain readily detects phosphoproteins containing phosphoserine, phosphothreonine, and phosphotyrosine residues on SDS-polyacrylamide gels, isoelectric focusing gels, 2-D gels, electroblots, and protein microarrays by a mechanism that combines a chelating fluorophore and a transition metal ion.[49–51] The staining is relatively rapid, simple to perform, readily reversible, and fully compatible with modern microchemical analysis procedures, such as MALDI–TOF–MS. Pro-Q Diamond dye can detect as little as 8 ng of pepsin, a monophosphorylated protein, and 1 ng of proteins with two or more phosphate residues, such as ovalbumin and β-casein.[51] The linear response of the fluorescent dye allows rigorous quantitation of phosphorylation changes over a 2–3 order of magnitude concentration range. Detection of phosphoproteins separated in IEF gels using Pro-Q Diamond dye initially presented certain challenges relative to standard SDS-polyacrylamide gels.[49] As found for other staining methods employed for detecting proteins in IEF gels, staining of phosphoproteins with Pro-Q Diamond dye was readily achieved after precipitating the proteins in 10% trichloroacetic acid/40% methanol and eluting the ampholytes by extensive washing.

VIII. ACQUIRING IMAGES FROM STAINED GELS

The introduction of fluorescent dye technology in particular has played a crucial role in the development of advanced imaging instrumentation.[52–54] The most common imaging instruments for protein visualization from electrophoresis gels use either a gas discharge transilluminator and charge-coupled device (CCD) camera or a photomultiplier tube and laser scanner. CCD cameras (14- or 16-bit cooled) are usually employed in CCD camera-based gel imagers, providing excellent quantitative information over a concentration range of 3 to 4

9 CAPILLARY ISOELECTRIC FOCUSING

TIM WEHR

Life Science Group, Bio-Rad Laboratories, 6000 James Watson Drive, Hercules, CA 94574

I. INTRODUCTION

The practice of isoelectric focusing in the capillary format provides the high resolving power of conventional gel isoelectric focusing (IEF) and the automation capabilities of instrumental techniques such as capillary electrophoresis (CE) and high-performance liquid chromatography (HPLC). The principle of capillary isoelectric focusing (cIEF) is similar to

Handbook of Isoelectric Focusing and Proteomics
D. Garfin and S. Ahuja, editors.

that of gel IEF: proteins migrate within a stable pH gradient formed by carrier ampholytes under the influence of an electric field. Upon attainment of equilibrium, proteins become focused within the pH gradient at points where they have zero net charge, i.e., their isoelectric points (pI). Any diffusion of the focused protein away from its isoelectric zone will result in acquisition of charge, resulting in back migration to the zone. The use of a narrow-bore fused silica capillary as the separation chamber provides efficient dissipation of Joule heat, enabling the use of very high electric fields (typically several hundred to a thousand V/cm). This allows separations to be performed in free solution, without the requirement for a gel as an anticonvective medium. The application of high field strengths provides high resolution (typically 0.02 pI units) and rapid analysis time. All steps in the analysis including introduction of sample and ampholytes, focusing, and protein detection can be performed automatically under instrument control, and the capillary can be reused for several hundred analyses. The ability to automate the IEF process and obtain quantitative information on resolved proteins is a driving force for replacement of gel IEF by cIEF, particularly in industrial settings.

A major limitation of performing IEF in capillaries is the use of fixed-point optical detectors on most commercial CE systems. In this approach, a short section of the capillary serves as the "flow cell" for on-tube detection. This is quite satisfactory for kinetic techniques such as capillary zone electrophoresis (CZE), but in cIEF it requires a means of transporting focused protein zones through the detection point without loss of resolution. A variety of techniques have been developed for mobilizing focused proteins which can be used singly or in concert. The added complexity of the mobilization process has been an obstacle in achieving reproducible cIEF separations. This chapter will describe the considerations for successful performance of cIEF. Several reviews of cIEF have been published,[1–5] so only recent applications and possible future developments will be discussed.

A CE system configured for cIEF is illustrated in Figure 1. The separation is carried out in a fused silica capillary which is coated externally with polyimide to provide mechanical strength. A portion of this polymer coating is removed at the far end of the capillary to serve as the detection "window." Most commercial CE systems employ UV absorbance detectors, but laser-induced fluorescence detection is occasionally used in cIEF. In many applications, the internal surface of the capillary is coated to suppress electro-osmotic flow (EOF). This phenomenon occurs when an electric field is applied to a capillary which is filled with an electrolyte and which possesses a fixed charge on the capillary wall. This causes flow of liquid toward the electrode with polarity opposite to that of the wall charge. In many CE applications, EOF is desirable since it can serve as a transport mechanism to carry all analytes to the detector. In most cIEF applications, EOF reduces

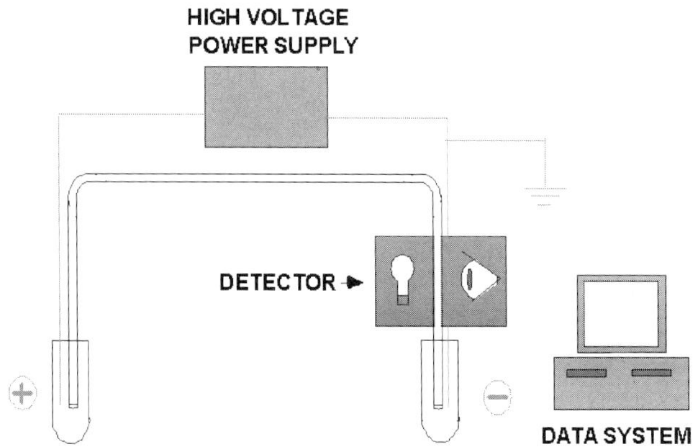

FIGURE 1 **Schematic diagram of a CE system configured for cIEF.**

resolution and reproducibility, so wall-coated capillaries are often preferred for cIEF. In a typical cIEF analysis, the capillary is first filled with a mixture of the sample and carrier ampholytes. This is accomplished by positioning a vial containing the mixture at the inlet side of the capillary with the capillary tip and high-voltage electrode immersed in the solution. The solution is forced into the capillary, typically by applying gas pressure to the head space above the sample vial or by applying vacuum to the outlet side of the capillary. Once the sample + ampholyte mixture is loaded into the capillary, the sample vial is replaced with the one containing anolyte (dilute acid), and a vial containing catholyte (dilute base) is positioned at the outlet end of the capillary. The second step in the cIEF process is application of high voltage to the capillary to initiate focusing. As focusing progresses, carrier ampholytes migrate to form a pH gradient and proteins migrate to their isoelectric points within the gradient. The final step in the cIEF process is mobilization of the capillary contents through the detector window. During this step, the detector signal is recorded to generate a profile of the focused proteins, similar to an HPLC chromatogram or the electropherogram obtained from a CZE experiment. Each of these steps in the cIEF process will be described in detail in the following discussion.

II. SAMPLE PREPARATION

Sample preparation for cIEF includes adjustment of sample ionic strength and protein concentration. These two parameters are the key to good performance in cIEF. Excessive sample ionic strength can have two negative consequences in the IEF process. First, high salt or buffer concentrations can cause excessive Joule heat during the initial stages of focusing, which can increase the risk of protein denaturation and precipitation. Second, the high-mobility salt ions exit the capillary during focusing, to be replaced by anolyte protons at one end of the capillary and catholyte hydroxyl ions at the other. Therefore, the pH gradient formed by the carrier ampholytes will be compressed in proportion to the amount of salt in the sample. This compression of the pH gradient can compromise resolution and induce protein precipitation. Also, exposure of the terminal ends of the capillary to extremes of pH, particularly the cathodic end, can reduce the longevity of the capillary when wall-coated capillaries are used.

The protein concentration of the sample can also affect the outcome of the cIEF experiment. If protein concentration is too low, detection sensitivity will be compromised and if it is too high, the risk of protein precipitation is increased. Precipitation of proteins occurs because the solubility of proteins under isoelectric conditions is reduced and the protein concentration in the focused zone can be elevated as much as 200-fold relative to the protein concentration in the initial sample.

A good rule of thumb is to adjust protein concentration to about 0.5 mg/mL and adjust sample ionic strength to 50 mM or lower. However, some proteins such as immunoglobulins, membrane proteins, and other large hydrophobic proteins are more prone to precipitation, and lower protein concentrations may be necessary for successful cIEF analysis. Filtration or centrifugation of the sample to remove any particulates or protein aggregates is also a good practice.

Off-line desalting techniques such as dialysis, ultrafiltration, or gel filtration are all satisfactory for preparing samples for cIEF. However, these techniques can be time consuming and laborious. Two on-line desalting techniques for cIEF have been described. Liao and Zhang[6] used an ampholyte-replacement procedure for desalting. In this approach, an acidic ampholyte solution titrated to pH 4 was used as anolyte, and an alkaline ampholyte solution titrated to pH 11 was used as catholyte. During the desalting step, salts present in the protein sample were exchanged for the titrated ampholytes. Following desalting, the titrated ampholyte solutions were replaced with conventional anolyte and catholyte solutions and cIEF was performed as usual. A limitation which affects reproducibility of this on-line desalting method is variation in the ampholyte distributions in the final pH gradients depending on the salt concentration of the sample.

A simpler on-line technique using voltage ramping has been described by Clarke et al.[7] At the beginning of focusing, capillary voltage was increased from 0 to 10 kV over several minutes. During this period, salt ions exited the capillary under low-voltage conditions, which minimized generation of Joule heat. This method is limited by the need to optimize voltage ramping conditions for different salt concentrations. Moreover, the problems of gradient compression and exposure of the capillary to elevated pH are not resolved with this method.

III. AMPHOLYTE SELECTION AND SAMPLE INTRODUCTION

The quality of the cIEF separation will depend upon the range and complexity of the carrier ampholyte mixture. For separation of proteins with a wide range of isoelectric points or to screen an unknown sample, a broad-range mixture of ampholytes is recommended, e.g., pH 3–10. For high resolution across a narrow pH range, a narrow-range ampholyte mixture can be used; there are several commercial sources for narrow-range mixtures that generate gradients spanning 1–3 pH units. However, narrow-range ampholyte preparations should be "doped" with a small amount (0.2–0.4% overall) of wide-range ampholytes to bridge the pH gap between the termini of the gradient and the anolyte and catholyte solutions.

The resolving power of an ampholyte pH gradient will depend on the number of ampholyte species in the mixture: the greater the number of ampholytes, the smaller will be the pH variance between adjacent loci within the capillary. Two approaches to increasing local resolution in cIEF have been described. Hjertén[8] proposed blending ampholyte mixtures from several suppliers to increase the number of ampholyte species. Righetti et al.[9] suggested adding specific zwitterionic species to the ampholyte blend to solve particular resolution problems. For example, Righetti et al. demonstrated that addition of β-alanine to a pH 6–8 ampholyte mixture could resolve hemoglobins A and F, and a combination of β-alanine and 6-aminocaproic acid added to the same ampholyte mixture could resolve hemoglobins A and A_{1C}.

When cIEF is performed in CE systems using on-tube detection, proteins which focus in the segment between the detection window and the capillary outlet may not be detected in some mobilization schemes. To confine the pH gradient within the "effective" length of the capillary (i.e., the length between the capillary inlet and the detection point), a spacer may be added to the sample + ampholyte mixture. A commonly used spacer for this purpose is *N,N,N′,N′*-tetramethylethylenediamine (TEMED).[10,11] At the completion of focusing, the TEMED spacer occupies the end of the capillary distal to the inlet, and all proteins focused within the pH gradient are detected. The appropriate amount of spacer can be determined from the ampholyte concentration and the percentage of the detector-distal capillary distance relative to the total capillary length. For example, when using an ampholyte concentration of 2%, a 20 cm capillary, and a 5 cm detection window from the capillary outlet, the appropriate TEMED concentration would be $5/20 \times 2\%$ or 0.5%.

An unfortunate feature of commercial ampholytes is their high background absorbance in the low-UV region. These products were all developed for conventional gel IEF, where UV absorbance is not an issue. cIEF with on-tube detection requires monitoring at longer wavelengths where the ampholytes are transparent. For most applications, detection at 280 nm is satisfactory. Although protein absorbance at 280 nm is typically

50–100-fold lower than that in the low UV region, the high protein concentration in the focused zones compensates for this loss in signal.

A novel solution to the problem of ampholyte background absorbance has been proposed by Huang et al.[12] These investigators performed cIEF in capillaries filled with pure water. The protons and hydroxyl ions formed by electrolysis of water at the high-voltage electrodes served to form a pH gradient within the capillary. However, the technique is limited by poor resolution and the tendency of proteins to precipitate in solutions of low ionic strength.

In most cIEF methods, the sample is premixed with the ampholyte mixture (and spacer, if appropriate) and the sample + ampholyte mixture is introduced into the capillary by the application of pressure at the capillary inlet, by application of vacuum at the capillary outlet, or by hydrodynamic injection. In the latter approach, the vial containing the sample + ampholyte mixture is elevated relative to the outlet vial. In contrast to CZE, a significant fraction of the capillary can be loaded with the sample, and in many approaches, the entire capillary is filled with the sample + ampholyte mixture. The large injection volumes used in cIEF result in increased zone concentrations and allow cIEF to be considered for micropreparative applications.

An alternative method for sample introduction in cIEF has been reported by Chen et al.[13] In their approach, the capillary was prefilled with carrier ampholytes, and the sample was introduced by electrokinetic injection. During injection, carrier ampholytes migrated to form a pH gradient, and analytes migrated to their isoelectric points within the gradient. The amount of sample loaded into the capillary was a function of the electric field strength and injection time. Enhancement of sample loading by 8–45-fold relative to the conventional cIEF sample introduction technique was demonstrated. A limitation of this dynamic sample introduction method is the injection bias of electrokinetic loading (analytes with low electrophoretic mobilities are injected with lower efficiencies than high-mobility analytes).

IV. FOCUSING

Once the capillary has been filled with the solution of carrier ampholytes, spacer, and proteins, focusing is initiated by application of high voltage. During this process, ampholytes and proteins are contained within the capillary using an acidic anolyte solution (typically 10–20 mM phosphoric acid) and an alkaline catholytic solution (typically sodium hydroxide at twice the anolyte concentration, e.g., 20–40 mM). It is recommended that the catholyte solution be freshly prepared to minimize the uptake of atmospheric carbon dioxide. The presence of carbonate salts in the catholyte can interfere with the focusing process. A typical field strength for focusing is 600 V/cm. Focusing is usually complete within a few minutes in short (15–20 cm) capillaries.

In the initial stages of focusing, high currents are typically observed as salts and buffer components in the sample migrate through the capillary. As the salts exit the capillary, and as ampholytes and proteins migrate to their isoelectric points, the capillary becomes depleted of current carriers and the observed current will drop exponentially. The attainment of equilibrium is evidenced by a drop in current to about 10% of the initial value and the rate of change of current approaches zero. This signals the completion of focusing, and continued application of high voltage can increase the risk of protein precipitation and loss of ampholytes.

At the onset of focusing, nascent protein zones form at both margins of the capillary and, as the final pH gradient is established, the zones coalesce at the isoelectric point of the protein. During this process, the nascent zones forming at the detection end of the capillary migrate through the detection window during their transit to their equilibrium positions. This "focusing electropherogram" can be used to monitor focusing, and can serve as a useful diagnostic tool (e.g., to detect capillary degradation and changes in EOF). Under normal conditions, the focusing electropherogram is very reproducible. It can be used to obtain analytical information about the sample when only a quick profile is needed,[14] or in cases where extended exposure of the sample to high voltage causes precipitation.

V. MOBILIZATION TECHNIQUES

The final step in cIEF, and the one that is unique to performing IEF in the capillary format, is mobilization of focused zones through the detection point. This requires application of a force to the capillary contents; this force can be electrophoretic, hydraulic (pressure, vacuum, or gravity), or electro-osmotic. It can be applied as a separate step following the completion of focusing (two-step cIEF) or applied during the focusing process (single-step cIEF). In all cases, high voltage is applied during mobilization to maintain zones in their focused state.

A. Two-step cIEF

Two-step cIEF is often the preferred approach since the focusing and mobilization conditions can be optimized independently.Two-step cIEF has been performed using electrophoretic and hydraulic mobilization.

1. Electrophoretic Mobilization

Electrophoretic mobilization has also been termed ion-addition mobilization or chemical mobilization. In all cases, it induces a shift in the charge state of the focused zones to cause them to move by electrophoretic

strongly at 280 nm, and cover a pH range 5.3–10.4. Recently, the same group has developed a series of fluorescent pI markers for use with laser-induced fluorescence detection in cIEF.[31] The markers are water-soluble derivatives of fluorescein containing phenolic and aliphatic amino groups. Four markers were described with pI values equal to 5.4, 5.7, 6.0, and 6.6.

IX. IMAGING cIEF

The mobilization process required for cIEF with conventional CE instruments with single-point on-tube detection has several limitations. Analysis time is increased using a separate mobilization stage (two-step cIEF) or using a reduced mobilization rate to allow complete focusing (single-step cIEF). Resolution may be compromised, particularly for proteins at the late-migrating terminus of the pH gradient. In EOF-driven mobilization, variations in mobilization velocity can compromise resolution and reproducibility. In hydraulic mobilization, laminar-flow band broadening can reduce resolution.

Imaging cIEF is a novel approach that eliminates the mobilization step. It was developed by Pawliszyn et al.[32,33] and a commercial imaging cIEF instrument has recently been introduced by Convergent Bioscience, Ltd. (Toronto, Ontario).[34,35] The system consists of a short (5 cm) capillary with the protective outer polyimide cladding removed, and the internal capillary coated with a hydrophilic polymer (Figure 6). Each end of

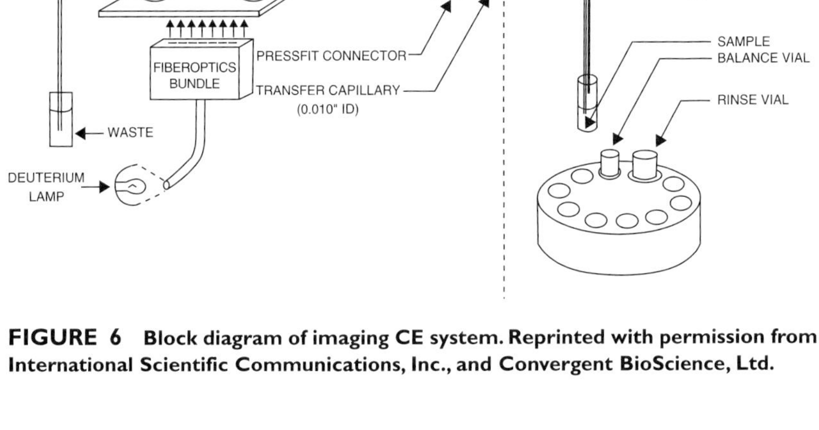

FIGURE 6 **Block diagram of imaging CE system. Reprinted with permission from International Scientific Communications, Inc., and Convergent BioScience, Ltd.**

the capillary is attached to sections of hollow-fiber dialysis tubing contained in the electrolyte reservoirs. These tubing segments serve to isolate the capillary contents from the electrolyte reservoirs but allow free passage of anolyte and catholyte ions.The sample contained in the sample loop of an eight-port injection valve is introduced into the capillary by an infusion pump. Injection can be performed manually or by using an autosampler. Following injection, high voltage is applied to initiate focusing. Detection is accomplished by illuminating the entire capillary with light from a xenon lamp delivered using a fiber optic system. Transmitted light from the entire capillary length is imaged onto a CCD detector. The focusing process can be monitored in real time for the development and optimization of the method, and the final zone profile at the completion of focusing can be captured for analytical purposes. Precipitation problems can be recognized by real-time monitoring, and the conventional solutions applied (addition of surfactants, chaotropes, organic modifiers, etc.). One advantage of imaging cIEF is that all segments of the pH gradient are focused for the same period, in contrast to conventional cIEF in which late-migrating proteins are focused for longer periods and are at greater risk for precipitation. The commercial imaging cIEF system can be equipped with an optional on-line desalting system for samples containing up to 150 mM salt. Imaging cIEF has recently been used to characterize glycoforms of recombinant human necrosis factor receptor FC fusion protein.[36]

X. cIEF-MASS SPECTROMETRY

On-line coupling of cIEF to mass spectrometry (MS) has generated interest in the growing field of proteomics because it shares some separation characteristics with 2-D gel electrophoresis (2-DE), a core technology in proteomics studies. In expression proteomics experiments, differences in the expression levels of cellular proteins in response to metabolic changes are used to identify proteins associated with disease progression, and to elucidate targets for therapeutic intervention. Such studies require the ability to detect quantitative changes in low-abundance proteins against a background of the thousands of proteins within a cell or tissue. To separate mixtures of such extraordinary complexity, a two-dimensional (2-D) separation technique is necessary. The ideal 2-D technique should have different selectivities in each dimension so that the total resolution of the technique is the product of the band capacity of the two dimensions. 2-DE fulfills this requirement since the first dimension (IEF) is based on charge (isoelectric point) while the second dimension (SDS–PAGE) is based on mass. Thus, the two dimensions are orthogonal in selectivity and 2-DE can resolve over 2000 proteins in a single gel. Unfortunately, 2-DE is time consuming, laborious, and only semi-quantitative. The coupling of cIEF to MS promised to provide the desired combination of orthogonal separation selectivities with automated analysis and short run times. The technique has been previously reviewed[37] and recent developments will be described here.

Successful coupling of cIEF with electrospray MS requires a means of performing focusing and mobilization with the capillary outlet interfaced with the ESI system. It also requires a means of preventing entry of the carrier ampholytes into the ionization system, since they can suppress analyte signal and foul the mass spectrometer. Initial attempts employed two-step cIEF with the catholyte reservoir placed in the ionization source during focusing.[38] Upon completion of focusing, the reservoir was removed and mobilization was initiated by infusion of a coaxial water methanol acetic acid sheath liquid. It was observed that the presence of ampholytes reduced protein net charge and ion intensities. In a later modification of this approach, gravity-assisted mobilization was used to compensate for moving-boundary effects caused by electromigration of sheath liquid ions into the capillary.[39] This cIEF–MS interface was coupled to a triple quadrupole MS for the analysis of transferrin glycoforms,[39] recombinant fusion proteins expressed in *Escherichia coli*[40] and phosphorylated albumins.[41] The same interface was coupled to a time-of-flight (TOF) mass spectrometer for analysis of model proteins.[42] Clarke and Naylor[43] described a two-step cIEF system in which only the composition of the sheath liquid was changed to effect mobilization. During focusing, the sheath liquid was methanolic ammonium hydroxide delivered at a reduced flow rate. This enabled formation of a static "hanging drop" of catholyte at the end of the separation capillary. At the completion of focusing, the sheath liquid was changed to anolyte (methanolic acetic acid) delivered at a higher flow rate to initiate electrophoretic mobilization into the ESI source of a double-sector mass spectrometer.

Problems encountered with replacement of the catholyte reservoir after focusing and with introduction of ampholytes into the mass spectrometer were solved by the use of microdialysis systems to remove ampholytes. Lamoree et al.[44] devised an on-line microdialysis (MD) system between the separation capillary and a transfer capillary connected to an ESI-quadrupole MS. The MD system consisted of hollow fiber dialysis tubing sealed in a chamber infused with acetic acid, which served as the catholyte. After focusing, contents of the separation capillary were mobilized through the MD using pressure, and ampholytes were removed by dialysis across the MD tubing. An acetic acid sheath liquid provided electrical contact for the ESI source. A later modification of this system replaced the hollow fiber dialysis tubing with a flat dialysis membrane.[45] The same group has also employed a free-flow electrophoresis (FFE) device to remove ampholytes prior to introduction of separated proteins into the ESI-MS.[46] In this device, an acetic acid carrier solution (which also served as the catholyte) was introduced into the FFE cell. A transverse electric field applied to the FFE cell caused a reduction in concentration of ampholytes by their differential deflection from the outlet of the cell. Focusing and mobilization were accomplished simultaneously by applying pressure at the capillary inlet in conjunction with a counterbalancing pressure supplied by the FFE carrier. A cIEF–MS system employing a hollow fiber microdialysis system and pressure mobilization was described by Yang et al.[47] In this system, anodic mobilization was used, and the microdialysis liquid (10% acetic acid) served as the anolyte and a proton source for ionization.

The extraordinarily high resolution and mass accuracy of Fourier transform ion cyclotron resonance MS (FTICR-MS) have made it a highly desirable technique in proteomic studies, since it enables unambiguous determination of protein mass and charge from a single charge state. Coupling of cIEF with FTICR-MS has been accomplished using the same approach described above for cIEF-quadrupole MS using two-step cIEF.[48,49] Focusing was carried out with a conventional setup using anolyte and catholyte reservoirs. After the focusing step, the capillary was inserted into the ESI source and focused zones mobilized by pressure in conjunction with water : acetic acid : methanol sheath flow. This approach has been used to identify carbonic anhydrase in a cell lysate in the presence of a 100-fold excess of hemoglobin[48] and to resolve ~900 proteins from the 3–60 kDa fraction of the *E. coli* proteome.[49] In the latter study, sensitivity and mass accuracy in the FTICR-MS analysis were improved by culturing cells in media depleted of the rare isotopes ^{13}C, ^{15}N, and ^{2}H.

XI. cIEF IN MICROCHANNELS

Chip-based microanalysis devices are gaining widespread attention for their promise in miniaturizing analytical instruments and providing platforms for high-throughput clinical diagnostics and rapid screening in drug discovery environments. A common format for these devices is the use of microchannel systems with electrically driven fluid transport. Several investigators have evaluated IEF in microchannel devices because it provides both high resolution and good sensitivity due to the high protein concentration of focused zones.

IEF in microchannels was employed by Hofmann et al.[50] as a detection method for fluorescent peptides used as probes in a multiplex diagnostic assay system. Single-step cIEF with EOF mobilization was used because of its compatibility with the chip format. The microchannel device consisted of a 200 μm × 10 μm channel of 7 cm length etched in planar glass. The Cy-5 labeled peptides were detected by laser-induced fluorescence.

form associated with human diabetes, has also been analyzed by cIEF.[65,66] cIEF has used to characterize the binding of hemoglobin to haptoglobin,[67] a serum glycoprotein which has a strong affinity for the α-globin chain.

B. Protein Glycoforms

Many proteins exist as isoforms in which oligosaccharide groups are attached to one or multiple sites on the primary sequence. Glycoforms can vary in the number of glycosylation sites occupied, and in the structure of the oligosaccharide at a particular site. The saccharide component of a glycoprotein can play a role in protein solubility, stability, and function; an understanding of protein glycosylation is therefore important in the development and manufacture of protein therapeutics. The presence of oligosaccharide groups can cause shifts of protein isoelectric points, so cIEF is an obvious tool for studying protein glycoforms. cIEF has been used to characterize glycoforms of recombinant tissue plasminogen activator,[68–71] erythropoietin,[72,73] recombinant human tissue necrosis factor receptor: FC fusion protein,[36] transferrins,[74–76] conalbumin,[75,77] metallothioneins,[75] and HIV envelope glycoproteins.[78] Techniques for separation of glycoproteins by cIEF are reviewed by Krull et al.[79]

C. Monoclonal Antibodies

Monoclonal antibodies, which are widely used as diagnostic and therapeutic tools, often exhibit microheterogeneity. These species arise from post-translational modifications such as glycosylation, or from alterations such as deamidation, clipping, and oxidation during purification, formulation and storage. cIEF is increasingly used to detect such microheterogeneity. Examples include monitoring production of recombinant antithrombin III,[80] quality control of humanized monoclonal HER2,[81] analysis of proteolysis fragments of humanized murine monoclonal antibodies,[82] characterization of monoclonal antibody isoforms,[83] and detection of charge heterogeneity in Mab C2B8 (a chimeric mouse/human monoclonal antibody directed to the human CD20 antigen).[84]

D. Peptides

Analysis of peptides by gel IEF is problematical because of their high diffusion rates, which causes loss of resolution when high voltage is turned off at the completion of focusing. Also, peptides have poor staining affinity. Although neither of these problems are relevant in cIEF, detection by on-tube UV detection is hampered by the high background absorbance of carrier ampholytes at the short wavelengths typically used for peptides. This necessitates detection at wavelengths of 280 nm or greater; therefore, only peptides with aromatic residues will be detected. Nonetheless, the high concentration factors in cIEF compensate for the reduced signal at higher wavelengths, and the high peak capacities (500–1000) have enabled cIEF to be used for separation of complex peptide mixtures. cIEF with UV detection has been applied to the separation of tryptic digests of model proteins[85–87] and yeast cytosolic proteins.[88] Mao and Zhang[89] used cIEF with laser-induced fluorescence to analyze BSA trypic peptides labeled with FITC or BODIPY. Cruikshank et al.[90] devised a detection scheme for nucleic acid hybridization probes using cIEF. Oligonucleotide probes were coupled via disulfide bonds to fluorescently tagged signal peptides. Following hybridization, the peptide tags were released with cysteine and separated by cIEF. A family of signal peptides, each with a distinct pI value, was used in a multiple-probe cIEF–LIF detection system.

E. Affinity cIEF

cIEF has occasionally been used to study bioaffinity interactions such as ligand–receptor binding. Okun[91] used cIEF to study the binding of actinavidin to biotin and to biotinylated oligonucleotides. It was observed that binding of the affinity ligand reduced the number of protein isoforms, and that the affinity complex exhibited a reduced pI. Righetti et al.[67] investigated the binding of haptoglobin to hemoglobin by cIEF. In this study, binding stoichiometry was determined by introducing haptoglobin into prefocused hemoglobin zones and detecting the acidic complexes as they migrated out of the pH gradient.

Lyubarskaya et al.[92] used cIEF-ion trap MS to determine affinity binding of tyrosine-phosphorylated peptides with the *src* SH2 domain, a non-catalytic region of a variety cellular proteins with tyrosine kinase activity. The concentrating power of the cIEF step provided increased sensitivity, while the MS^n capabilities of the ion trap MS provided structural information for ligand identification.

F. cIEF in Proteomics

As discussed earlier, the enormous complexity of protein extracts from cells and tissues presents a formidable challenge in protein identification and quantitation in proteomic studies. With a resolving power better than 0.01 pI and a concentrating power better than 500-fold, cIEF may represent the most powerful single-dimension separation technique for proteins.[93] The application of cIEF in proteomics has been the subject of several reviews.[5,93–96]

The high resolution attainable with cIEF alone makes it suitable for separation of protein mixtures of moderate complexity. For example,

Manabe et al. were able to separate about 60 proteins in human plasma using two-step cIEF with electrophoretic mobilization.[97,98] However, for the samples commonly encountered in proteomics studies, a single-dimension separation technique is inadequate, and multidimensional approaches are more widely used. This can be done off-line, and cIEF has been used as the second-dimension separation after capillary reversed-phase chromatography.[89] However, on-line multidimensional approaches are preferred for considerations of throughput and automation. Liquid-phase separations such as cIEF are easily coupled to electrospray ionization interfaces, so cIEF–ESI–MS has been explored as an on-line 2-D analytical system for proteomics. Jensen et al.[99] were able to resolve 400–1000 proteins in the mass range 2–100 kDa in cell lysates of *E. coli* and *Deinococcus radiodurans* using cIEF coupled on-line with FTICR–MS. Shen et al.[88] demonstrated cIEF separation of the peptides in a tryptic digest of yeast cytosol proteins; resolution of ~0.005 pI units produced an estimated peak capacity of ~1000.

A novel 2-D capillary separation system coupling cIEF with transient isotachophoresis (ITP) and CZE was developed by Mohan and Lee.[87] The two separation modes were joined by a microdialysis device containing acetic acid, which served as the anolyte for the first-dimension cIEF separation and the background electrolyte for the second-dimension ITP/CZE separation. Following focusing in the cIEF dimension, segments of the pH gradient were hydrodynamically injected into the ITP/CZE capillary by gravity. The carrier ampholytes served as the leading electrolyte and acetic acid as the terminating electrolyte for transient ITP. Since the bulk of the carrier ampholyte population migrated ahead of the analytes, detection in the low-UV region was possible. The authors demonstrated a 2-D separation of tryptic peptides obtained from a three-component protein mixture. The two separation modes are largely orthogonal in separation mechanism, and the peak capacity of ~1600 was estimated for the 2-D peptide separation.

Sheng and Pawliszyn[86] have used the imaging cIEF system described above as the second dimension in a 2-D system. CZE was used as the first dimension for separation of model proteins, and micellar electrokinetic chromatography (MEKC) was used as the first dimension for separation of tryptic peptides. In both cases, EOF was used to transport the electrolyte from the first-dimension capillary. A ten-port valve carrying two hollow-fiber dialysis loops served as the interface between the two separation systems. The dialysis loops functioned to remove electrolytes and detergents from the first-dimension eluent, and to introduce carrier ampholytes and urea for the second-dimension separation. An additional eight-port valve was used to flush the contents of the imaging cIEF system between duty cycles. Alternating capture of first-dimension eluent in one loop and second-dimension analysis of the contents of the other loop permitted continuous on-line sample analysis. An advantage of using imaging cIEF in the second dimension is that diffusional band broadening in the interface was eliminated by the focusing power of cIEF. The distribution of peptide "spots" across the entire surface of a 2-D presentation of the MEKC–cIEF data demonstrated the orthogonality of the two separation systems.

G. Other Applications

cIEF has been used for a number of novel applications. These include separation and detection of protein complexes,[100,101] separation of microorganisms,[102,103] and analysis of organic selenium complexes by cIEF–ICP–MS.[104]

REFERENCES

1. Wehr, T., Zhu, M. and Rodriguez-Diaz, R. Capillary isoelectric focusing. In *Methods in Enzymology* Vol. 270 (Karger, B. L. and Hancock, W. S. Eds.) Academic Press, San Diego, pp. 358–374, 1996.
2. Rodriguez-Diaz, R., Wehr, T., Zhu, M. and Levi, V. Capillary isoelectric focusing. In *Handbook of Capillary Electrophoresis*, 2nd Ed. (Landers, J. Ed.) CRC Press, Boca Raton, pp. 101–153, 1996.
3. Wehr, T., Rodriguez-Diaz, R. and Zhu, M. Capillary isoelectric focusing. In *Capillary Electrophoresis of Proteins*, *J. Chromatogr. Library*, Vol. 80, Marcel Dekker, New York, pp. 131–233, 1998.
4. Wehr, T., Rodriguez-Diaz, R. and Zhu, M. Recent advances in capillary isoelectric focusing. In *CE in Biotechnology: Practical Applications for Protein and Peptide Analysis* (Altria, K. D. Ed.) *Chromatographia CE Series*, Vol. 53, Vieweg Publishing, Weisbaden, pp. 45–58, 2001.
5. Shimura, K. Recent advances in capillary isoelectric focusing: 1997–2002. *Electrophoresis 2002* 23:3847–3857, 2002.
6. Liao, J. L. and Zhang, R. Simple approach to eliminating disturbances in isoelectric focusing caused by the presence of salts. *J. Chromatogr.* 684:143–148, 1994.
7. Clarke, N. J., Tomlinson, A. J., Schomberg, G. and Naylor, S. Capillary isoelectric focusing of physiologically derived proteins with on-line desalting of isotonic salt concentrations. *Anal. Chem.* 69:2786–2792, 1997.
8. Hjertén, S. Isoelectric focusing in capillaries. In *Capillary Electrophoresis: Theory and Practice* (Grossman, P. D. and Colburn, J. C. Eds.) Academic Press, San Diego, pp. 191–214, 1992.
9. Righetti, P. G., Bossi, A. and Gelfi, C. Capillary isoelectric focusing and isoelectric buffers: an evolving scenario. *J. Capillary Electrophoresis* 4:47–59, 1997.
10. Zhu, M., Rodriguez, R. and Wehr, T. Optimizing separation parameters in capillary isoelectric focusing. *J. Chromatogr.* 559:479–488, 1991.
11. Mohan, D. and Lee, C. S. Extension of separation range in capillary isoelectric focusing for resolving highly basic biomolecules. *J. Chromatogr. A* 979:271–276, 2002.
12. Huang, T., Wu, X.-Z. and Pawliszyn, J. Capillary isoelectric focusing without carrier ampholytes. *Anal. Chem.* 72:4758–4761, 2000.
13. Chen, J., Gao, J. and Lee, C. S. Dynamic enhancements of sample loading and analyte concentration in capillary isoelectric focusing for proteome studies. *J. Proteome Res.* 2:249–254, 2003.

14. Kilar, F. and Hjertén, S. Fast and high resolution analysis of human serum transferrin by high performance isoelectric focusing in capillaries. *Electrophoresis* 10:23–29, 1989.
15. Hjertén, S. and Zhu, M. Adaptation of the equipment for high-performance electrophoresis to isoelectric focusing. *J. Chromatogr.* 346:265–270, 1985.
16. Hjertén, S., Liao, J.-L. and Yao, K. Theoretical and experimental study of high-performance electrophoretic mobilization of isoelectrically focused protein zones. *J. Chromatogr.* 387:127–138, 1987.
17. Zhu, M., Rodriguez, R. and Wehr, T. Optimizing separation parameters in capillary isoelectric focusing. *J. Chromatogr.* 559:479-488, 1991.
18. Hjertén, S. High-performance electrophoresis. Elimination of electroendosmosis and solute adsorption. *J. Chromatogr.* 347:191–198, 1985.
19. Chen, S.-M. and Wiktorowicz, J. E. Isoelectric focusing by free solution capillary electrophoresis. *Anal. Biochem.* 206:84–90, 1992.
20. Mazzeo, J. R. and Krull, I. S. Capillary isoelectric focusing of proteins in uncoated fused-silica capillaries using polymeric additives. *Anal. Chem.* 63:2852–2857, 1991.
21. Mazzeo, J. R. and Krull, I. S. Coated capillaries and additives for the separation of proteins by capillary zone electrophoresis and capillary isoelectric focusing. *Biotechniques* 10:638–645, 1991.
22. Thormann, W., Caslavska, J., Molteni, S. and Chmelik, J. Capillary isoelectric focusing with electroosmotic zone displacement and on-column multichannel detection. *J. Chromatogr.* 589:321–327, 1992.
23. Whynot, D. M., Hartwick, R. A. and Bane, S. Capillary isoelectric focusing with anionic coated capillaries. *J. Chromatogr.* 767:231–239, 1997.
24. Kilár, F., Végvári, Á. and Mód, A. New set-up for capillary isoelectric focusing in uncoated capillaries. *J. Chromatogr.* 813:349–360, 1998.
25. Wehr, T. Recent advances in capillary electrophoresis columns. *LC-GC.* 11:14–20, 1993.
26. Conti, M., Galassi, M., Bossi, A. and Righetti, P. G. Capillary isoelectric focusing: the problem of protein solubility. *J. Chromatogr.* 757:237–245, 1997.
27. Kobayashi, A. H., Aoiki, M., Suzuki, M., Yanagisawa, A. and Arai, E. Evaluation of pH gradient formation of carrier ampholytes with synthesized isoelectric point markers in capillary isoelectric focusing. *J. Chromatogr. A* 772:137–144, 1997.
28. Shimura, K., Wang, Z., Matsumoto, H. and Kasai, K. Synthetic oligopeptides as isoelectric point markers for capillary isoelectric focusing with ultraviolet absorption detection. *Electrophoresis 2000* 21:603–610, 2000.
29. Shimura, K., Kamiya, K., Matsumoto, H. and Kasai, I. Fluorescence-labeled peptide pI markers for capillary isoelectric focusing. *Anal. Chem.* 74:1046–1053, 2002.
30. Šlais, K. and Friedl, Z. Low-molecular-mass pI markers for isoelectric focusing. *J. Chromatogr. A* 661:249–256, 1994.
31. Šlais, K., Horká, M., Nováčková, J. and Friedl, Z. Fluorescein-based pI markers for capillary isoelectric focusing with laser-induced fluorescence detection. *Electrophoresis 2002* 23:1682–1688, 2002.
32. Wu, J. and Pawlyszyn, J. A capillary cartridge with an on-line desalting device that allows fast sampling for capillary isoelectric focusing. *Anal. Chem.* 67:2010–2014, 1995.
33. Fang, X., Tragas, C., Wu, J., Mao, Q. and Pawlyszyn, J. Recent developments in capillary isoelectric focusing with whole-column imaging detection. *Electrophoresis* 19:2290–2295, 1998.
34. Wu, J. and Watson, A. Automated sample introduction for an imaged capillary isoelectric focusing instrument via high-performance liquid chromatography sampling devices. *J. Chromatogr. B* 714:113–118, 1998.
35. Wu, J., Li, S.-C. and Watson, A. Optimizing separation conditions for proteins and peptides using imaged capillary isoelectric focusing. *J. Chromatogr. A* 817:163–171, 1998.
36. Jochheim, C., Novick, S., Balland, A., Mahan-Boyce, J., Wang, W.-C., Goetze, A. and Gombotz, W. Separation of Enbrel (rhuTNFR:Fc) isoforms by capillary isoelectric focusing. In *CE in Biotechnology: Practical Applications for Protein and Peptide*

Analysis (Altria, K. D. Ed.) *Chromatographia CE Series* Vol. 53, Vieweg Publishing, Weisbaden, pp. 59–65, 2001.

37. Banks, J. F. Recent advances in capillary electrophoresis/electrospray mass spectrometry. *Electrophoresis* 18:2255–2266, 1997.
38. Tang, Q., Harrata, A. K. and Lee, C. S. Capillary isoelectric focusing-electrospray mass spectrometry for protein analysis. *Anal. Chem.* 67:3515–3519, 1995.
39. Yang, L., Tang, Q., Harrata, A. K. and Lee, C. S. Capillary isoelectric focusing-electrospray ionization mass spectrometry for transferrin glycoforms analysis. *Anal. Biochem.* 243:140–149, 1996.
40. Tang, Q., Harrata, A. K. and Lee, C. S. Two-dimension analysis of recombinant *E. coli* proteins using capillary isoelectric focusing electrospray ionization mass spectrometry. *Anal. Chem.* 69:3177–3182, 1997.
41. Wei, J., Yang, L. Harrata, A. K. and Lee, C. S. High resolution analysis of protein phosphorylation using capillary isoelectric focusing-electrospray ionization-mass spectrometry. *Electrophoresis* 19:2356–2360, 1998.
42. Wei, J., Lee, C. S., Lazar, I. M. and Lee, M. L. Capillary isoelectric focusing-electrospray ionization time-of-flight mass spectrometry for protein analysis. *J. Microcol. Sep.* 11:193–197, 1999.
43. Clarke, N. and Naylor, S. Capillary isoelectric focusing-mass spectrometry: analysis of protein mixtures from human body fluids. *Biomed. Chromatogr.* 16:287–297, 2002.
44. Lamoree, M. H., Tjaden, U. R. and van der Greef, J. Use of microdialysis for the online coupling of capillary isoelectric focusing with electrospray mass spectrometry, *J. Chromatogr. A.* 777:31–39, 1997.
45. Lamoree, M. H., van der Hoeven, R. A. M., Tjaden, U. R. and van der Greefr, J. Application of microdialysis for on-line coupling of capillarly isoelectric focusing with electrospray mass spectrometry on a magnetic sector instrument, *J. Mass Spectrom.* 33:453–460, 1998.
46. Chartogne, A., Tjaden, U. R. and van der Greef, J. A free-flow electrophoresis chip device for interfacing capillary isoelectric focusing on-liine with electrospray mass spectrometry. *Rapid. Commun. Mass Spectrom.* 14:1269–1274, 2000.
47. Yang, L., Lee, C. S., Hofstadler, S. A. and Smith, R. D. Characterization of microdialysis acidification for capillary isoelectric focusing-microelectrospray ionization mass spectrometry. *Anal. Chem.* 70:4945–4950, 1998.
48. Severs, J. C., Hofstadler, S. A., Zhao, Z., Senh, R. T. and Smith, R. D. The interface of capillary electrophoresis with high performance Fourier transform ion cyclotron resonance mass spectrometry for biomolecule characterization. *Electrophoresis* 17:1808–1817, 1996.
49. Jensen, P. K., Paša-Tolic, L., Anderson, G. A., Horner, J. A., Lipton, M. S., Bruce, J. E. and Smith, R. D. Probing proteomes using capillary isoelectric focusing-electrospray ionization Fourier transform ion cyclotron resonance mass spectrometry. *Anal. Chem.* 71:2076–2084, 1999.
50. Hofmann, O., Che, D., Cruickshank, K. A. and Müller, U. R. Adaptation of capillary isoelectric focusing to microchannels on a glass chip. *Anal. Chem.* 71:678–686, 1999.
51. Rossier, J. S., Schwarz, A., Reymond, F., Ferrigno, R., Bianchi, F. and Girault, H. H. Microchannel networks for electrophoretic separations. *Electrophoresis* 20:727–731, 1999.
52. Tan, W., Fan, Z. H., Qiu, C., Ricco, A. J. and Gibbons, I. Miniaturized capillary isoelectric focusing in plastic microfluidic devices. *Electrophoresis 2002* 23:3638–3645, 2002.
53. Tsai, S.-W., Loughran, M., Hiratsuka, A., Yano, K. and Karube, I. Application of plasma-polymerized films for isoelectric focusing of proteins in a capillary electrophoresis chip. *Analyst* 128:237–244, 2003.
54. Herr, A. E., Molho, J. I., Drouvalakis, K. A., Mikkelsen, J. C., Utz, P. J., Santiago, J. G. and Kenny, T. W. On-chip coupling of isoelectric focusing and free solution electrophoresis for multidimensional separations. *Anal. Chem.* 75:1180–1187, 2003.

55. Bolger, C. A., Zhu, M., Rodriguez, R. and Wehr, T. Performance of uncoated and coated capillaries in free zone electrophoresis and isoelectric focusing of proteins. *J. Liq. Chromatogr.* 14:895–906, 1991.
56. Zhu, M., Wehr, T., Levi, V., Rodriguez, R., Shiffer, K. and Cao, Z. A. Capillary electrophoresis of abnormal hemoglobins associated with α-thalassemias. *J. Chromatogr. A* 652:119–129, 1993.
57. Zhu, M., Rodriguez, R., Wehr, T. and Siebert, C. Capillary electrophoresis of hemoglobins and globin chains. *J. Chromatogr.* 608:225–237, 1992.
58. Molteni, S., Frischknecht, H. and Thormann, W. Application of dynamic capillary isoelectric focusing to the analysis of human hemoglobin variants. *Electrophoresis* 15:22–30, 1994.
59. Hempe, J. M. and Craver, R. D. Quantification of hemoglobin variants by capillary isoelectric focusing. *Clin. Chem.* 40:2288–2295, 1994.
60. Mario, N., Baudin, B., Aussel, C. and Giboudeau, J. Capillary isoelectric focusing and high-performance cation exchange chromatography compared for qualitative and quantitative analysis of hemoglobin variants. *Clin Chem.* 43:2137–2142, 1997.
61. Hempe, J. M., Granger, J. N., Warrier, R. P. and Craver, R. D. Analysis of hemoglobin variants by capillary isoelectric focusing. *J. Capillary Electrophoresis* 4:131–135, 1997.
62. Mohammad, A. A., Okorodudu, A. O., Bissell, M. G., Dow, P., Reger, G., Meier, A., Guodagno, P. and Petersen, J. R. Clinical application of capillary isoelectric focusing on fused silica capillary for determination of hemoglobin variants. *Clin. Chem.* 43:1798–1799, 1997.
63. Mario, N., Baudin, B. and Giboudeau, J. J. Qualitative and quantitative analysis of hemoglobin variants by capillary isoelectric focusing. *J. Chromatogr. B*, 706:123–129, 1998.
64. Jenkins, M. A. and Ratnaike, S. Capillary isoelectric focusing of haemoglobin variants in the clinical laboratory. *Clin. Chim. Acta* 289:121–132, 1999.
65. Conti, C., Gelfi, A., Bosisio, B. and Righetti, P. G. Quantitation of glycated hemoglobins in human adult blood by capillary isoelectric focusing. *Electrophoresis* 17:1590–1596, 1996.
66. Hempe, J. M. and Craver, R. D. Quantification of hemoglobin variants by capillary isoelectric focusing. *Clin. Chem.* 40:2288–2295, 1994.
67. Righetti, P. G., Conti, M. and Gelfi, C. Study of haptoglobin-hemoglobin complexes by titration curves, capillary electrophoresis and capillary isoelectric focusing *J. Chromatogr. A* 767:255–262, 1997.
68. Yim, K. W. Fractionation of the human recombinant tissue plasminogen activator (rtPA) glycoforms by high-performance capillary zone electrophoresis and capillary isoelectric focusing. *J. Chromatogr.* 559:401–410, 1991.
69. Thorne, J. M., Goetzinger, W. K., Chen, A. B., Moorhouse, K. G. and Karger, B. L. Examination of capillary zone electrophoresis, capillary isoelectric focusing and sodium dodecyl sulfate capillary electrophoresis for the analysis of recombinant tissue plasminogen activator. *J. Chromatogr. A* 744:155–165, 1996.
70. Kubach, J. and Grimm, R. Non-native capillary isoelectric focusing for the analysis of the microheterogeneity of glycoproteins. *J. Chromatogr. A* 737:281–289, 1996.
71. Moorhouse, K. G., Eusebio, C. A., Hunt, G. and Chen, A. B. Rapid one-step capillary isoelectric focusing method to monitor charged glycoforms of recombinant human tissue-type plasminogen activator. *J. Chromatogr. A* 717:61–69,1995.
72. Cifuentes, A., Moreno-Arribas, M. V., de Frutos, M. and Diez-Masa, J. C. Capillary isoelectric focusing of erythropoietin glycoforms and its comparison with flat-bed isoelectric focusing and capillary zone electrophoresis. *J. Chromatogr. A* 830:453–463, 1999.
73. Lopez-Soto-Yarritu, P., Diez-Masa, J. C., Cifuentes, A. and de Frutos, M. Improved capillary isoelectric focusing method for recombinant erythropoietin analysis. *J. Chromatogr. A* 968:221–228, 2002.
74. Kilar, F. and Hjerten, S. Fast and high resolution analysis of human serum transferrrin by high performance isoelectric focusing in capillaries. *Electrophoresis* 10:23–29, 1989.

75. Richards, M. P. and Huang, T. L. Metalloprotein analysis by capillary isoelectric focusing. *J. Chromatogr. B* 690, 43–54, 1997.
76. Wu, J. and Pawliszyn. *In vitro* observation of interactions of iron and transferrin by capillary isoelectric focusing with a concentration gradient imaging detection system. *J. Chromatogr. A* 652:295–299, 1993.
77. Huang, T. L. and Richards, M. P. Development of a high-performance capillary isoelectric focusing technique with application to studies of microheterogeneity in chicken conalbumin. *J. Chromatogr. A* 757:247–253, 1997.
78. Tran, N. T., Taverna, M., Chevalier, M. and Ferrier, D. One-step capillary isoelectric focusing for the separation of the recombinant human immunodeficiency virus envelope glycoprotein glycoforms. *J. Chromatogr. A* 866:121–135, 2000.
79. Krull, S., Kazmi, S., Zhong, H. and Santora, L. C. Separation of glycoproteins by capillary isoelectric focusing. In *Methods in Molecular Biology*,Vol. 213 (Thibault, P. and Honda, S. Eds.) Humana Press, Totowa, pp. 197–218, 2003, NJ.
80. Reif, O.-S. and Freitag, R. Control of the cultivation process of antithrombin III and its characterization by capillary electrophoresis. *J. Chromatogr. A* 680:383–394, 1994.
81. Hunt, G., Moorhouse, K. G. and Chen, A. B. Capillary isoelectric focusing and sodium dodecyl sulfate-capillary gel electrophoresis of recombinant humanized monoclonal antibody HER2. *J. Chromatgr. A* 744:295–301, 1996.
82. Hagmann, M. L., Kionka, C., Schreiner, M. and Schwer, C. Characterization of the $F(ab')_2$ fragment of a murine monoclonal antibody using capillary isoelectric focusing and electrospray ionization mass spectrometry. *J. Chromatogr. A* 816:49–58, 1998.
83. Lee, H. G. Rapid high-performance capillary isoelectric focusing of monoclonal antibodies in uncoated fused-silica capillaries. *J. Chromatogr. A* 790:215–223, 1997.
84. Hunt, G., Hotaling, T. and Chen, A. B. Validation of a capillary isoelectric focusing method for the recombinant monoclonal antibody C2B8. *J. Chromatogr. A* 800: 355–367,1998.
85. Mazzeo, J. R., Martineau, J. A and Krull, I. S. Peptide mapping using EOF-driven capillary isoelectric focusing. *Anal. Biochem.* 208:323–329, 1993.
86. Sheng, L. and Pawliszyn, J. Comprehensive two dimensional separation based on coupling micellary electrokinetic chromatography with capillary isoelectric focusing. *Analyst* 127:1159–1163, 2002.
87. Mohan, D. and Lee, C. S. On-line coupling of capillary isoelectric focusing with transient isotachophoresis-zone electrophoresis: A two-dimensional separation system for proteomics. *Electrophoresis 2002* 23:3160–3167, 2002.
88. Shen, Y., Berger, S. J., Anderson, G. A. and Smith, R. D. High-efficiency capillary isoelectric focusing of peptides. *Anal. Chem.* 72:2154–2159,2000.
89. Mao, Y. and Zhang, X. Comprehensive two-dimensional separation system by coupling capillary reversed-phase liquid chromatography to capillary isoelectric focusing for peptide and protein mapping with laser-induced fluorescence detection. *Electrophoresis 2003* 24:3289–3295, 2003.
90. Cruickshank, K. A., Olvera, J. and Muller, U. R. Simultaneous multiple analyte detection using fluorescent peptides and capillary isoelectric focusing. *J. Chromatogr. A* 817:41–47, 1998.
91. Okun, V. Affinity mode of capillary isoelectric focusing for the characterization of the biotin-binding protein actinavidin. *Electrophoresis* 19:427–432, 1998.
92. Lyubarskaya, Y. V., Carr, S. A., Dunnington, D., Prichett, W. P., Fisher, S. M., Appelbaum, E. R., Jones, C. S. and Karger, B. L. Screening for high-affinity ligands to the *Src* SH2 domain using capillary isoelectric focusing-electrospray ionization ion trap mass spectrometry. *Anal. Chem.* 70:4761–4770, 1998.
93. Shen, Y. and Smith, R. D. Proteomics based on high-efficiency capillary separations. *Electrophoresis 2002* 23:3106–3124, 2002.
94. Manabe, T. Capillary electrophoresis of proteins for proteomic studies. *Electrophoresis 1999* 20:3116–3121,1999.

95. Manabe, T. Combination of electrophoretic techniques for comprehensive analysis of complex protein systems. *Electrophoresis 2000* 21:1116–1122, 2000.
96. Manabe, T. Analysis of complex protein-polypeptide systems for proteomic studies. *J. Chromatogr. B* 787:29–41, 2003.
97. Manabe, T., Miyamoto, H. and Iwasaki, A. Effects of catholytes on the mobilization of proteins after capillary isoelectric focusing. *Electrophoresis* 18:92–97, 1997.
98. Manabe, T., Iwasaki, A. and Miyamoto, H. Separation of human plasma/serum proteins by capillary isoelectric focusing in the absence of denaturing agents. *Electrophoresis* 18:1159–1165, 1997.
99. Jensen, P. K., Paša-Tolić, L., Peden, K. K., Martinović, S., Lipton, M. S., Anderson, G. A., Tolic, N., Wong, K.-K. and Smith, R. D. Mass spectrometric detection for capillary isoelectric focusing separations of complex protein mixtures. *Electrophoresis* 21: 1372–1380, 2000.
100. Martinović, S., Berger, S. J., Paša-Tolić, L. and Smith, R. D. Separation and detection of intact noncovalent protein complexes from mixtures by on-line capillary isoelectric focusing-mass spectrometry. *Anal. Chem.* 72:5356–5360, 2000.
101. Shen, Y., Berger, S. J. and Smith, R. D. High-efficiency capillary isoelectric focusing of protein complexes from *Escherichia coli* cytosolic extracts. *J. Chromatogr. A* 914:257–264, 2001.
102. Horká, M., Planeta, J., Růzićka, F. and Šlais, K. Sol-gel column technology for capillary isoelectric focusing of microorganisms and biopolymers with UV or fluorometric detection. *Electrophoresis* 24:1383–1390, 2003.
103. Shen, Y., Berger, S. J. and Smith, R. D. Capillary isoelectric focusing of yeast cells. *Anal. Chem.* 72:4603–4607, 2000.
104. Michalke, B. and Schramel, P. Application of capillary zone electrophoresis-inductively coupled plasma mass spectrometry and capillary isoelectric focusing-inductively coupled plasma mass spectrometry for selenium speciation. *J. Chromatogr.* 807:71–80, 1998.

10

FREE-FLOW ISOELECTRIC FOCUSING

PETER J. A. WEBER, GERHARD WEBER, CHRISTOPH ECKERSKORN, ULRICH SCHNEIDER, AND ANTON POSCH

FFEWeber GmbH IZB, Building 6, Am Klopferspitz 19, D-82152 Planegg/Munich, Germany

I. INTRODUCTION

Free-flow electrophoresis (FFE) represents one of the most versatile preparative-scale fractionation and separation techniques used in (bio)chemistry.[1–4] It was first described more than 40 years ago and entitled "matrix-free, preparative flow-through electrophoresis"[5] and "matrix-free, continuous-flow electrophoresis (CFE)".[6] These descriptions

Handbook of Isoelectric Focusing and Proteomics
D. Garfin and S. Ahuja, editors.

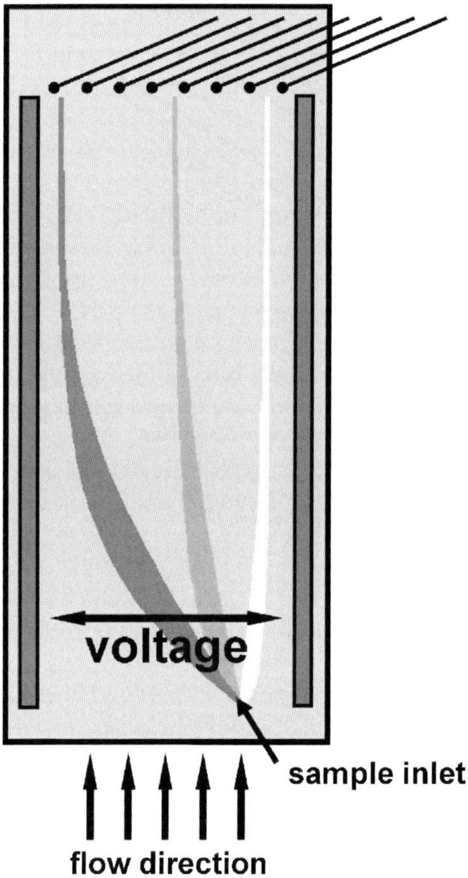

FIGURE 2 Separation scheme of FF-IEF experiments. After entering the separation chamber directly, the samples are relatively diffuse, because the pH gradient has not been formed yet. In parallel with the formation of the pH gradient, the samples become aligned according to their individual pI values. Their zones become sharper and sharper and remain sharp due to the focusing effect of the pH gradient.

separation buffers of constant composition, pH, and conductivity are used. This means that the sample components are merely separated according to their constant electrophoretic mobilities (charge to size ratio) at a given pH. Therefore, the sample injection beam has to be as narrow as possible, because there is no focusing effect to counteract band broadening. ZE is predominantly used for the separation of cells, membranes, and organelles, because they typically do not have discrete pI values. In addition, the ZE separation buffers are less complex and more flexible than the IEF separation buffers, thus allowing milder separation conditions.

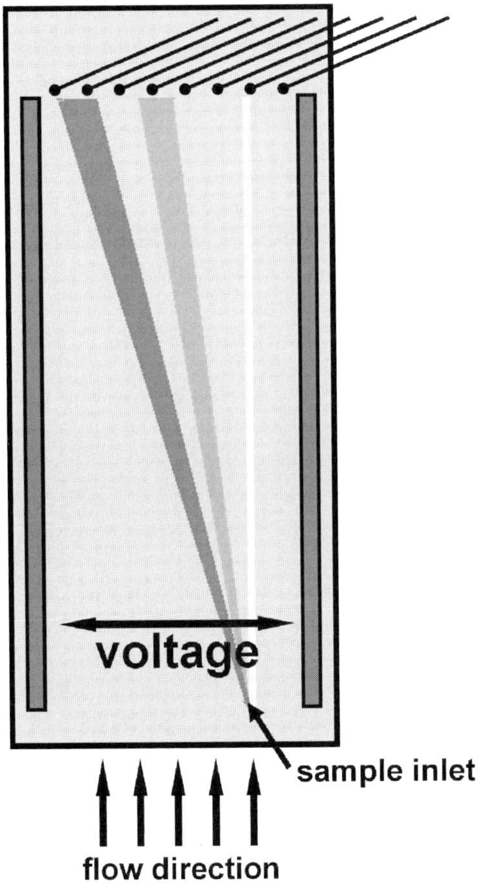

FIGURE 3 **Separation scheme of FF-ZE experiments. Since the typical FFE-related distortion effects are not counteracted by a focusing effect, the sample stream gets broader and broader on its way through the separation chamber. In addition, the absence of any focusing effect leads to a continual migration, perpendicular to the laminar flow.**

2. FSE

FSE is similar to ZE, except that the conductivity of the separation buffer is not uniform across the separation chamber: A low-conductivity buffer is pumped through the chamber adjacent to a high-conductivity one. The sample (typically dilute) is introduced into the chamber as a very broad sample beam via the low-conductivity buffer. As for ZE, the sample migrates towards the high-conductivity medium on its way through the chamber, driven by the electrical field. Since the voltage is inversely proportional to the conductivity, the sample will be retarded as soon as it reaches the conductivity step, i.e., the sample will focus at the interface between the high- and low-conductivity buffer. Thus, large amounts of sample (several g/h) can be concentrated to a sharp band and harvested with a high concentration into one fraction tube. Accordingly, FSE is very useful for the high-throughput concentration of dilute solutions of rather pure proteins.

3. ITP

Many compounds are insoluble in buffers having pH values similar to their pI. For them, FFE-ITP may be an alternative method. Generally, in ITP, at least three different buffers are necessary. The first buffer contains the so-called leading ions having a high electrophoretic mobility. The middle buffer is composed of the sample and so-called spacing ions having intermediate electrophoretic mobilities. The last buffer contains the so-called terminating ions having a low electrophoretic mobility. When an electric field is applied, the sample ions line up between the leading ions, the spacing ions, and the terminating ions according to their own electrophoretic mobilities. ITP is similar to the more familiar "stacking" phenomenon, but there is no "unstacking" mechanism, such as a sharp decrease in gel pore size. The quality of the separation highly depends on the sample components and on the choice of the spacing ions. Some components will be concentrated at boundaries between spacing ions because their electrophoretic mobilities fit between those of the spacers. Other components will be spread throughout the zones of the spacing ions with equal electrophoretic mobilities.

D. Important Theoretical Parameters

1. Hydrodynamic Distortion

FFE separation buffers show a non-turbulent streamline flow profile typical for liquids that flow as layers between two parallel plates. This so-called laminar flow profile is characterized by a parabolic shape, i.e., the fluid velocity is zero at all bounding walls and reaches a maximum midway between the walls. Laminar flow is an intrinsic problem of all kinds of FFE separations, because sample ions flowing midway between the walls will spend less time in the electric field than sample ions near the walls. This leads to broad sample bands having crescent shapes. The phenomenon, referred to as hydrodynamic distortion, can be reduced by either increasing the distance between the separation chamber walls or by reducing the size of the sample band at the injection point. However, the first measure will cause thermal distortions (see Section III.D.2) and the second measure will reduce the throughput.

2. Thermal Distortion

An inherent problem of all electrophoresis systems is the so-called Joule heating. This temperature increase occurs whenever a current passes through a conductor such as the FFE separation buffer. Because the chamber walls can dissipate the heat that is generated during FFE, the buffer near the walls will be cooler than the buffer midway between the walls. These temperature differences are equivalent to density differences and will cause thermal convections, which will distort the separation. Thus, it is a basic requirement to use narrow separation chambers (thickness <1.0 mm). Additional cooling of the separation buffer and/or the walls of the separation chamber is highly recommended. Based on theoretical considerations, there are a variety of other measures that prove useful to reduce thermal distortions; e.g., high flow rates reduce the time that the separation buffer spends in the electric field thus limiting system heating. Use of separation buffers with low conductivities, the application of low voltages, and narrower electrode cross sections also helps to reduce Joule heating. Convective thermal distortion, i.e., density differences, can be completely avoided by running the experiment in space at zero gravity (see Section III.A.2).

3. Electrodynamic Distortion

Charged double layers near the separation chamber walls caused by ionizable functional groups of the wall material very often result in a net flow close to the walls toward the cathode. This phenomenon, known as electro-osmotic flow, can be another reason for band broadening. Sample or buffer ions near the walls are dragged toward the cathode, whereas the ions midway between the walls remain unaffected by this surface effect. This so-called electrodynamic distortion leads to crescent shapes of sample bands similar to those caused by the hydrodynamic distortion. In cases where electrodynamic and hydrodynamic distortion oppose each other, the corresponding band broadening might vanish; in all other cases it will worsen the situation. To minimize the electro-osmotic flow, measures have to be taken to reduce the number of ionizable functional groups at the surfaces of the chamber walls. This can be achieved by the replacement of easily ionizable glass walls with appropriate plastic walls, by the static covalent modification of surfaces with appropriate polymers, or by the addition of surface-active compounds like hydroxypropylmethylcellulose (HPMC), to the separation buffer that will coat the surfaces dynamically.

4. Electrohydrodynamic Distortion

Conductivity differences between the sample and the separation buffer cause sheer stresses that are independent of the previous distortion phenomena.[23,24] This so-called electrohydrodynamic distortion leads to changes of the sample stream shape (flattening in the direction toward the wall or toward the electrodes), thus also causing band broadening. The phenomenon can be eliminated by careful adjustment of the sample conductivity and the separation buffer conductivity.

E. Important Practical Parameters

1. External Flow Distortion

The turbulence-free flow of the separation medium is easily distorted by external factors such as air bubbles or lint. To avoid the introduction of lint it normally suffices to use lint-free paper tissues for cleaning chamber walls and to filter the separation buffers before use. To prevent the introduction of air bubbles the liquid levels of all separation buffers have to be checked regularly. For FF-IEF experiments, stationary air bubbles in the first third of the separation chamber might be tolerable, because of the focusing principle that will make up for the distortion, but most of the time they lead to a permanent disturbance that requires the abortion of the experiment.

2. Reagent Purity

It is important to use only high-purity water as well as chemicals of the highest grade available. Ionic impurities dramatically increase the conductivity of the separation buffers or can be focused and precipitated in the separation chamber, which would lead to a massive distortion of the separation process. The quality differences of detergents are specifically pronounced, because the detergents themselves as well as their impurities interact not only with the sample components such as proteins, but also with the surfaces of the separation chamber walls as well leading to unpredictable phenomena.

3. Separation Buffer

Each separation exercise requires a suitable set of media. Important considerations in media selection include: (a) the electrolytes rinsing the electrodes must optimally transfer electrical power from the electrodes to the chamber without causing detrimental effects on the separation process or the sample; (b) the actual separation media in the chamber must favor separability, solubility, and activity of the compound of interest;[21,22,25] (c) the anodal and cathodal margin media flowing laterally in the separation chamber (if applicable) must prevent contamination of the membranes with compounds from the separation area and vice versa to protect the separation area from the influence of the electrodes (for FF–IEF this is typically achieved by addition of ions with a high electrophoretic mobility like sulphuric acid for the anode and sodium hydroxide for the cathode);[26] and (d) all media in the separation chamber must be adjusted with regard to their specific densities and viscosities at the temperature of the experiment to avoid unpredictable phase boundary effects and/or turbulences.

4. Sample Preparation

To avoid electrohydrodynamic distortions (see above) as well as unpredictable phase boundary effects and turbulences it is crucial to match the density, viscosity, and conductivity of the sample solution and the separation buffer. This can be easily achieved by preparation of a concentrated sample solution and its subsequent dilution with separation buffer (at least 1:5). If the sample components are not stable at the pH of the separation buffer, or the sample concentration would be too low after dilution with the separation buffer, solid buffer ingredients can be added directly to the sample to yield the same viscosity, density, and conductivity.

5. Sample Precipitation

Two factors increase the tendency of samples to precipitate during IEF experiments: the low solubility of compounds at their isoelectric point and the high local concentration of focused compounds. Very often, this leads to the precipitation of the one or a few compounds out of a mixture having the highest concentration and/or the lowest solubility at their pI values. If the FFE instrument design allows following the separation visually, precipitation can be easily detected as sharp white lines occurring in the separation chamber. As long as these lines stay sharp and keep moving, one can keep on fractionating the sample. When "flakes" (that might clog the fractionation outlets) or stationary "islands" (that might cause turbulences and local overheating) form, the sample flow has to be reduced, or the sample has to be diluted, or some detergent has to be added to the sample and/or separation buffer.

6. Sedimentation

There are two kinds of sedimentation: particle sedimentation and zone sedimentation. Particle sedimentation is a result of density differences between the sample (mainly in the case of cells and organelles) and the separation buffer. This most often occurs when the density of the sample particles is higher than the density of the surrounding separation buffer and the sample particles sink. This directly leads to band broadening. Particle sedimentation can be avoided by adjusting the density of the separation buffer by running the FFE experiment in the vertical position or in an exotic approach by running the experiment under microgravity (see Section III.A.2.). Zone sedimentation occurs when buffer ingredients or sample components accumulate in a narrow zone, e.g., by IEF. This effect might make this zone denser than the surrounding solution, causing the zone to sediment within the solution. This phenomenon is virtually impossible to avoid, except by running the experiment under microgravity (see Section III.A.2.).

F. Alternative Liquid-based IEF Techniques

There are a variety of alternative liquid-based methods and instruments for the pI-dependent separation of proteins (see Table 1). All of them depend on some kind of compartmentalization utilizing a membrane,

TABLE 1 Alternative Instruments for the pI-dependent Separation of Proteins

Instrument name	Manufacturer	Recent publications[a]
Gradiflow™ BF 400	Gradipore, Frenchs Forest, NSW, Australia	28–33
IsoelectrIQ²	Proteome Systems, North Ryde, NSW, Australia	34, 35
IsoPrime™	Amersham Biosciences, Buckinghamshire, UK	36
Off-Gel™ electrophoresis	DiagnoSwiss, Monthey, Switzerland	37, 38
Rotofor®	Bio-Rad, Hercules, CA, USA	39–44
ZOOM® IEF Fractionator	Invitrogen, Carlsbad, CA, USA	45–47

[a]Some publications discuss prototype instruments and others discuss applications.

grid, or matrix. This is a fundamental difference to FF-IEF, because any interaction on passage through a membrane, grid, or matrix inevitably suffers from a high risk of material loss. In addition, the compartment-based instruments are typically designed for batch-wise processing, which limits their throughput when compared with the continuous separation by FF-IEF. Nevertheless, they prove to be very useful for certain separation needs (see Table 1). Since they are not in the scope of this chapter, we refer those interested to the preceding chapter on "Alternative forms of IEF" and an excellent recent review.[27]

III. INSTRUMENTATION

A. Historical Overview

1. The Early Years

A few years after the publication of the basic FFE principles in the late 1950s and early 1960s of the last century,[5,6] the development of the first commercial FFE instruments started and within 10–20 years a huge variety of instruments was in the market. These early commercial devices included the FF machine from Brinkmann Instruments (Westbury, NY, USA);[48,49] the FF-4 instrument from Desaga (Heidelberg, Germany);[50] the FF-5 instrument (Biomedical Instruments, New York, NY, USA);[51] the ACE 710 (Hirschmann, Unterhaching, Germany);[52,53] the BIOSTREAM with an annulus geometry (CJB Developments, Portsmoth, USA);[54,55] and above all the VaP machines (VaP-5, VaP-11, VaP-21, VaP-22, VaP-220 of Bender & Hobein, Munich, Germany).[56–60] In particular, Hannig and his co-workers supported this maturation of the FFE technology, but others, like Strickler, made important contributions too. Hundreds of instruments were sold and were used in laboratories around the world for the successful separation of all kinds of compounds from low-molecular-weight organic compounds, peptides, and proteins, to organelles and whole cells.[10] This is documented by a multitude of publications in the 1960s, to 1980s.[61–63] However, most of these experiments were performed done in the ZE mode and to a lesser extent in the ITP or FSE mode. Of course, the high resolution achievable by IEF encouraged many to develop a free-flow version of the proces, too, but none of them was really successful.[64–68] This was mainly due to very long focusing times requiring either a slow flow through the chamber or a very long chamber, the uncontrollable pH drift toward the cathode (electro-osmosis), and the necessity to reduce the voltage/current levels because of insufficient heat dissipation (thermal convection). As a consequence, this led to two new trends: recycling instruments and compartmentalized instruments (mainly in combination with external coolers) or hybrids thereof. One of the most innovative scientist was Bier, who introduced the recycling isoelectric focusing (RIEF) instrument[69,70] and the commercially very successful Rotofor® (Bio-Rad, Hercules, CA, USA).[71] Righetti made very important contributions too, which led to the development of the membrane-based multicompartment IsoPrime™ system (Amersham Biosciences, Buckinghamshire, UK).[72] Other recycling instruments that entered the market at that time were the RF3™ by Protein Technologies (Tucson, AZ, USA), and the ATIsolator™ (Ampholife Technologies, The Woodlands, TX, USA).

2. FFE in Space

In the 1980s, space agencies like NASA had some difficulties justifying their expensive space missions. Thus, they sought beneficial scientific experiments that suffer from gravity on earth to help legitimize their flights, and to increase public acceptance. FFE offered a perfect system. After a variety of space experiments in the 1970s and early 1980s using static column electrophoresis, actual FFE experiments began in 1982 during space shuttle mission STS-4 using a McDonnell Douglas CFE system (CFES) and during the Soviet Salyut-7 mission using the FFE system "Tavriya". More than ten space-flight missions followed,[73] and several other nations like Germany (Messerschmitt-Bölkow-Blohm (MBB-ERNO, now DASA) with their TEM 06–13),[74] Japan (NASDA space agency and Mitsubishi Heavy Industries with their FFE unit FFEU),[75] and France (Matra with their free-flow device RAMSES)[76] did their share too. Comparing the results of all of these experiments and their impact on FFE technology on Earth with hundreds of millions of dollars that they cost, their outcome is absolutely meagre. Nevertheless, the Japanese FFE microgravity research continues to the present day.[77] An excellent review about FFE in space was published recently.[78]

3. The Recent Years

In the late 1980s, stagnation of FFE technology, particularly in the field of FF-IEF, the reduced interest in traditional FFE fields like organelle separations, the rise of new fields like molecular biology, and improvements in other techniques like RP-HPLC for peptides and small proteins and FACS for cells gradually ousted FFE and its manufacturers from the market. However, in the early 1990s, Dr. Weber GmbH (Ismaning, Germany) developed a new FFE instrument called OCTOPUS (see Figure 4) addressing the technological and experimental problems with a variety of innovative concepts.[79,80] They established a reproducible IEF mode with high resolution[15] and improved the general robustness of the FFE process by means of so-called "margin" and "counterflow" media.[26] Combined with an intelligent patent policy, this allowed the company to survive the ongoing selection and consolidation process mentioned above as the only provider of FFE instruments in the late 1990s. This was just about the time when the whole field of biological sample fractionation experienced a second spring with the advent of the proteomics era.[81] The reduction of sample complexity either at the protein or at the organelle level is a prerequisite for successful proteomic analyses. Consequently, there is a dramatic need for reproducible, versatile separation processes with high throughput and reasonable resolution like liquid-based IEF methods or liquid chromatography.[27,82] Dr. Weber GmbH was acquired by TECAN (Männedorf, Switzerland) in order to use FFE as the initial part of an automated platform for proteome analyses (see below). After a complete redesign of the OCTOPUS, the so-called Pro Team™ FFE (see Figure 5) offers a unique portfolio of

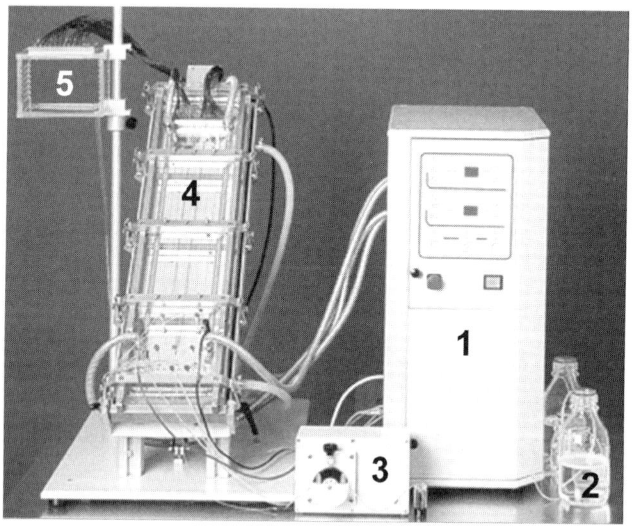

FIGURE 4 **OCOTPUS FFE instrument (Dr. Weber GmbH): 1, tower containing power supply, electrode pump, media pump, and electronic control unit; 2, separation media bottles; 3, sample pump; 4, separation chamber; 5, fractionation housing to hold 96-well plates.**

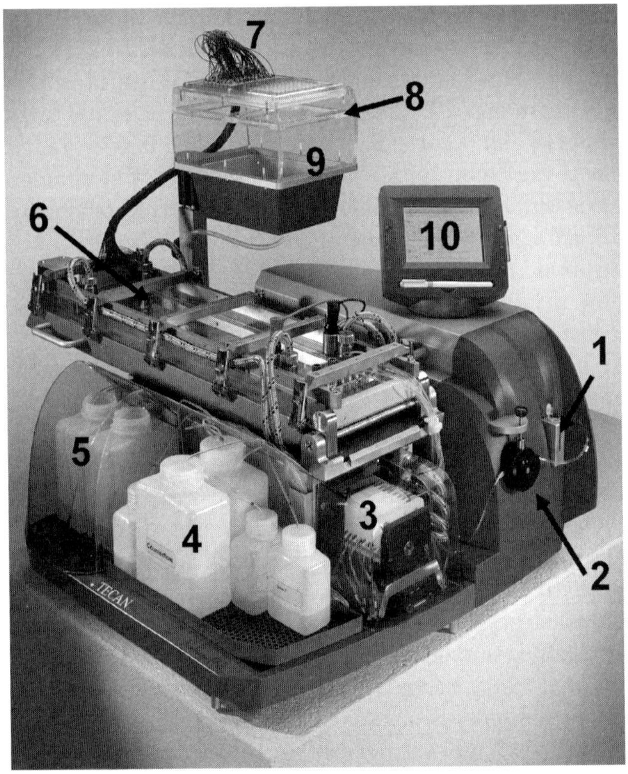

FIGURE 5 Pro Team™ FFE instrument (Tecan): 1, sample holder; 2, sample pump; 3, separation media pump; 4, separation media bottles; 5, electrolyte bottles; 6, separation chamber; 7, fractionation tubing; 8, 96-well plate holder; 9, waste; 10, touch screen control unit. The power supply is integrated in the system.

methods for the fractionation and prefractionation of proteins and organelles by IEF and ZE, respectively (see the section "Applications"). Mainly due to the patent situation it is still the only free-flow instrument in the market. Nevertheless, falling back on compartmentalized devices other companies sought to offer alternative liquid-based instruments for the pI-dependent separation of proteins. These devices include the Gradiflow™ BF 400 (Gradipore, Frenchs Forest, Australia), IsoelectrIQ^2 (Proteome Systems, North Ryde, Australia), Off-Gel™ electrophoresis (DiagnoSwiss, Monthey, Switzerland), and ZOOM® IEF Fractionator (Invitrogen, Carlsbad, CA, USA) (Table 1).

B. Pro Team™ FFE

Principally, an FFE instrument consists of a separation chamber, a fraction collector, a cooling device, a high-voltage power supply, several pumps

proteome: only the high-abundant proteins are visualized, due to the inability of current staining techniques to display the lowest abundance together with the highest abundance species on one gel.

One of the most promising strategies for increasing the total number of proteins that can be detected from complex proteomes is sample pre-fractionation prior to 2-DE. The central intention of all existing prefractionation techniques is to separate crude protein samples into a small number of well-resolved fractions. Four major pre-fractionation approaches have been described:

- cell fractionation by centrifugation
- sequential extraction according to differences in protein solubility
- chromatography (e.g., ion exchange, hydroxyapatite, and affinity resins)
- liquid-phase electrophoresis techniques

When using FFE for the prefractionation of complex protein samples the device is usually operated in the IEF mode thereby dividing the proteome into discrete pI ranges (Figure 13).

To compensate for dilution during the FF run and to remove additives that are not compatible with further downstream analytical techniques like 2-DE or chromatographic separations, an appropriate sample cleanup

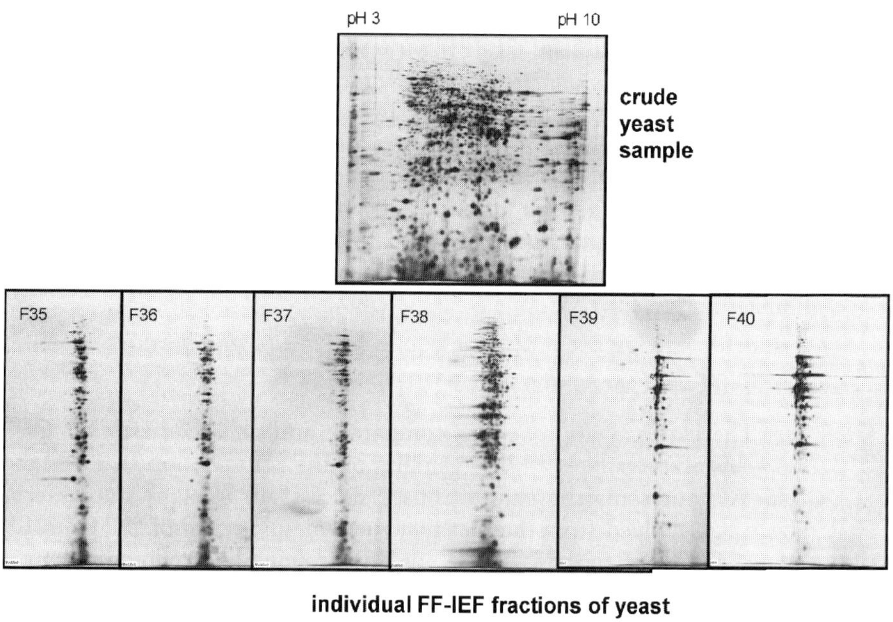

FIGURE 13 **2-D electrophoresis (pH 3–10) of a crude yeast protein sample (upper panel) and the same sample fractionated by FF-IEF (lower panel).**

and concentration step should be considered. Ultrafiltration is a well-established method to process protein samples in the desired way. Ultrafiltration of the FFE fractions can be performed by filtration devices, equipped with 96 cavities. The 96 cavities are arranged in a 12×8 microtiterplate format and the volume of each cavity is about 2 mL. Ultrafiltration membranes with different cutoffs, from 5 to 20 kDa are available. The FFE fractions can be collected into these devices prior ultrafiltration. FFE fractions with a volume of >2 mL can be transferred to single filtration devices and concentrated. Various filtration devices with volumes up to 25 mL are available. Ultrafiltration devices, manufactured by Vivascience, offer good recovery rates, even for membrane proteins.

To take advantage of the continuous-operation mode of FF, a 96-well on-line protein-binding device like a solid-phase-extraction (SPE) plate operated under vacuum conditions is also appropriate. The SPE plate may contain any type of resin that allows proteins to bind with high capacity such as butyl- (C_4-), ion exchange, Hydrophilic interaction (poly-hydroxyethyl aspartamide), hydrophobic interaction or MC resins. The characteristics of the chosen downstream analytical techniques must be compatible with the elution conditions and thus define the resin type that may be used in the SPE plate. For the powerful combination of FFE and 2-DE or HPLC, sample cleanup by solid-phase extraction utilizing poly-(2-hydroxyethyl)-aspartamide (poly-HEA) silica is suited best since elution can be conducted in aqueous buffer solution. Hydrophilic interaction chromatography (HILIC) is a long-standing variant of normal-phase chromatography, which binds proteins to a strongly hydrophilic support based on interaction of hydrophilic parts of the proteins with the resin. The method and its basic principles have been extensively described.[96] Applications of HILIC include the isolation of membrane proteins,[97] glycopeptides,[98] and post-translationally modified protein variants.[99]

In HILIC, binding occurs in highly concentrated organic solvents like acetonitrile or propanol, whereas elution is performed by flushing the resin with aqueous solutions such as 2-DE sample buffer or starting buffer for reversed-phase chromatography, which is usually 0.1% trifluoroacetic acid. Typically, samples from FFE experiments are diluted in a 10-fold excess of a high-organic content buffer to enable binding of the proteins to the resin. In this context, the counterflow is used to deliver the required amount of HILIC binding solution, thereby mixing the fractions with the binding solution directly at the site of fractionation (on-line sample processing). Figure 14 shows a prototype setup of such a processing device. The samples are dispensed automatically into a 96-well microtiter plate filled with approximately 100 mg of poly-HEA silica resin per well.

Figure 15 demonstrates the FFE-postprocessing power of HILIC applied to individual yeast fractions collected from a denaturing FF-IEF

FIGURE 14 **Photograph of a prototype setup of the FFE fractionation unit linked to a solid-phase extraction device. Ninety-six fractionation tubes lead into the upper orifices of a HILIC SPE plate, which is constantly held under light vacuum (approximately 350 mbar). The liquid migrating through the tubing consists of HILIC binding buffer and FF-IEF media containing focused proteins. Mixing of the two liquids occurs within the tubing and at the outlet.**

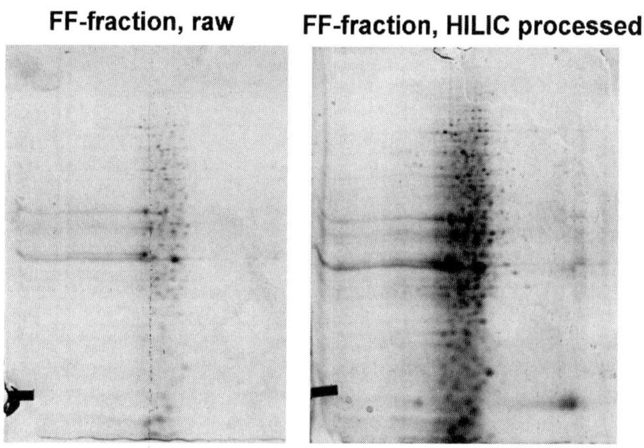

FIGURE 15 **FF-IEF 2-DE coupling. One fraction of an FF-IEF run with total yeast protein extract has been processed by HILIC sample concentration procedure. Left panel: 2-D gel (silver stained) of an individual FF-IEF fraction loaded without any further processing. Right panel: The same fraction after HILIC processing. Note that the loaded volumes were maintained equal in both experiments.**

run. The left panel shows the result of 2-DE without prior sample processing by HILIC, whereas the right panel shows the result of the very same fraction after sample processing by HILIC followed by compatible elution in 2-DE sample buffer. Detailed qualitative and quantitative computer-aided image analysis of both processed and unprocessed FF fractions, analyzed with 2-DE clearly indicates that no non-specific protein losses occurred during the HILIC manipulations. Moreover, the processed FF samples show clear protein enrichment while the overall quantitative proportions have been retained. The detection limits of 2-DE when combined with FF-IEF are further improved when HILIC processed fractions are separated on narrow-range pH gradients spanning 1 or 1.5 pH units in the first dimension of 2-DE. Approaching low-abundance proteins by increasing the protein load of crude, non-fractionated samples to 2-D gels, although narrow-range pH gradients are used is not sufficient, since resolution is quickly lost due to massive protein precipitation. On the other hand, by applying the protein fractions obtained by FF-IEF onto appropriate narrow-range IPG intervals, much higher sample loads can be used, since only the proteins co-focusing in the chosen pH gradient will be present (Figure 16).

The experiment shown in Figure 17 shows the utility of HILIC sample processing in the context of FF-IEF HPLC coupling: the upper panel shows a reversed-phase chromatogram from a fraction that has been loaded directly onto the column, the lower panel shows the same fraction after HILIC sample processing. The concentration effect achieved by HILIC sample processing becomes evident by comparing the two chromatograms that were generated from identical fractions. After sample processing, the overall intensity together with the resolution of individual peaks is much more pronounced, in fact, if FF-IEF fractions are injected directly onto the column, no chromatographic peaks are discernible.

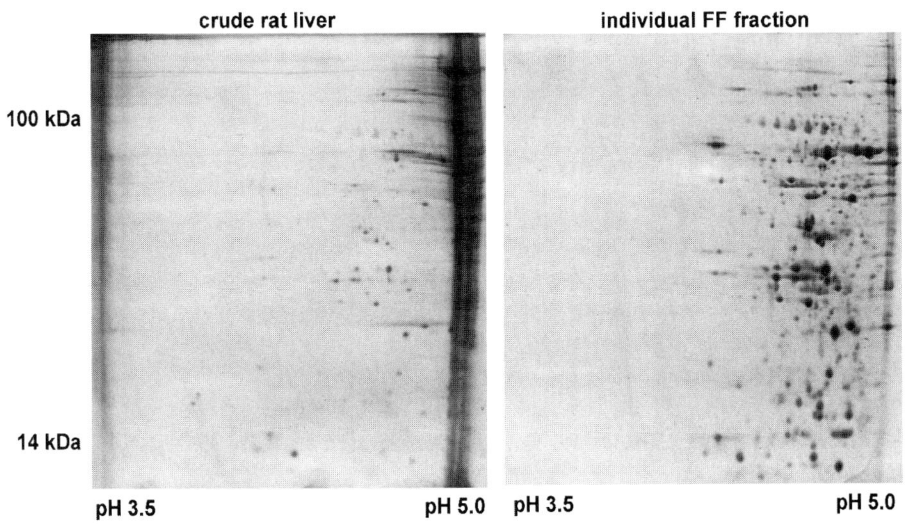

FIGURE 16 Rat liver proteins analyzed by 2-DE (pH 3.5–5.0): comparison of raw (left) and FF-IEF fractionated liver sample (right).

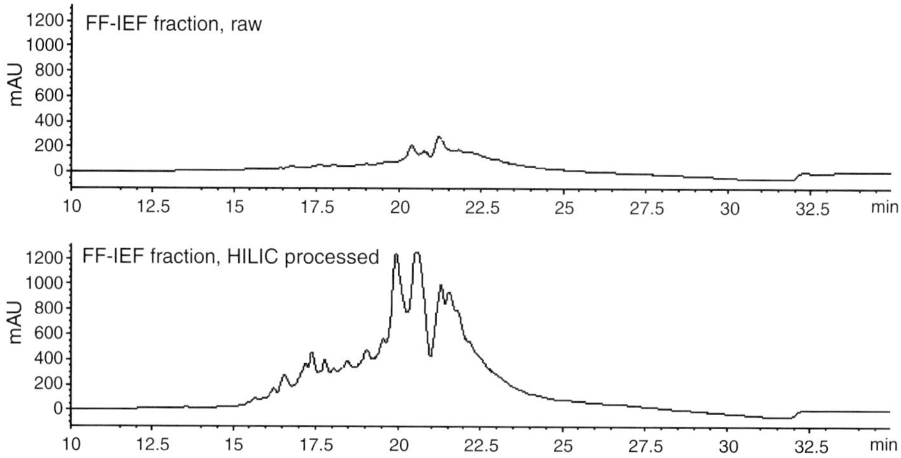

FIGURE 17 **FF-IEF HPLC coupling. An FFE fraction was injected onto a Poros R2/10 (Applied Biosystems GmbH, Darmstadt, Germany) polymeric reversed-phase column. The gradient solvents were: A, 0.1% trifluoroacetic acid; B, solvent A plus 80% acetonitrile. The gradient was developed from 100% solvent A and 80% solvent B over 60 min with 214 nm recording displayed. Upper panel: chromatogram of a fraction injected directly onto the column. Lower panel: the same fraction after HILIC processing as described. Note that the same volumes (500 µL) were injected in both experiments.**

These examples clearly demonstrate the value of combining HILIC and FF-IEF in enabling current proteomics technologies such as 2-DE or HPLC to display many more protein species than currently achievable without prefractionation.

D. Other FF-IEF Applications

The only FF-IEF results, apart from protein separations, that were published recently were from the Vigh lab and deal with the separation of low-molecular-weight enantiomers like dansyl-tryptophan.[100,101] These papers document the continuing efforts to optimize the purification of racemic mixtures based on their different interactions with enantiomerically pure additives like cyclodextrins.

E. Non-FF-IEF Applications

FFE modes other than IEF are not within the scope of this chapter, but might be of interest to some readers. Since the latest review dates back to 1998, we compile some of the most recent references. This should enable the reader to get a complete overview of the possibilities of FFE.

The ZE mode was mainly applied for the separation of organelles and membranes.[102–111] In addition, it was used for the isolation of a peptide,[112] for the characterization and fractionation of cells,[12,113] and the separation of chiral compounds.[114–117] FF-ITP was used for the separation of humic acids and metal humates[118] as well as the separation of rare-earth EDTA chelates.[119] An FSE–ZE-mixed mode was used for the purification of a protein.[20]

V. SUMMARY

Free-flow electrophoresis (FFE), also known as continuous free-film electrophoresis or continuous-flow electrophoresis (CFE) is one of the most versatile preparative-scale fractionation and separation techniques available in (bio)chemistry. FFE utilizes a thin film of separation buffer that flows continuously in a laminar fashion between two narrowly spaced plates and an electric field that is applied perpendicular to the flow. FFE results in a differential deflection of charged samples as they move toward the collection ports. This allows the high-throughput separation of all types of samples such as low-molecular-weight organic compounds, peptides, proteins, protein complexes, membranes, organelles, and whole cells. FFE supports all modes of electrophoresis such as zone electrophoresis (ZE), field-step electrophoresis (FSE), isotachophoresis (ITP), and isoelectric focusing (IEF). In this review, we focus on FF-IEF and attempt to provide the reader with a comprehensive overview of its principles covering all relevant parameters, the historical, state of the art, and future instrumentation as well as the most recent applications. Wherever appropriate, we additionally include general information about FFE for completeness. Furthermore, we touch on related technologies such as multi-compartment electrolyzers (MCE) to allow their proper differentiation.

REFERENCES

1. Roman, M. C. and Brown, P. R. Free-flow electrophoresis. *Anal. Chem.* 66(2):86–94, 1994.
2. Krivankova, L. and Bocek, P. Continuous free-flow electrophoresis. *Electrophoresis* 19(7):1064–1074, 1998.
3. Bauer, J. Advances in cell separation: recent developments in counterflow centrifugal elutriation and continuous flow cell separation. *J. Chromatogr. B* 722(1–2):55–69, 1999.
4. Pasquali, C., Fialka, I. and Huber, L. A. Subcellular fractionation, electromigration analysis and mapping of organelles. *J. Chromatogr. B* 722:89–102, 1999.
5. Barrollier, J., Watzke, E. and Gibian, H. Einfache Apparatur für die trägerfreie präparative Durchlauf-Elektrophorese. *Z. Naturforschung.* 13b:754–755, 1958.

51. Sachs, G. et al. A nonelectrogenic H^+ pump in plasma membranes of hog stomach. *J. Biol. Chem.* 251(23):7690–7698, 1976.
52. Eckstein, R. et al. Der Einsatz eines trägerfreien Elektrophoresesystems (ACE 710) zur Früherkennung einer positiven Reaktion in der gemischten Lymphozytenkultur. In *Elektrophoreseforum '83*, (Radola, B. J. Ed.) TU München, pp. 139–146, 1983.
53. Bauer, J. and Kachel, V. Separation accuracy of free flow electrophoresis as proved by flow cytometry. *Electrophoresis* 9(1):62–63, 1988.
54. Wenger, P. et al. Purification of lentil lectins using preparative electrophoresis. *J. Chromatogr.* 455:225–239, 1988.
55. Rutte, R. Free-flow electrophoresis as a purification step of high versatility for bioproducts. *Ber. Bunsen Phys. Chem.* 93:1038–1042, 1989.
56. Harms, E., Kern, H. and Schneider, J. A. Human lysosomes can be purified from diploid skin fibroblasts by free-flow electrophoresis. *Proc. Natl. Acad. Sci.* USA 77(10): 6139–6143, 1980.
57. Pesonen, M., Ansorge, W. and Simons, K. Transcytosis of the G protein of vesicular stomatitis virus after implantation into the apical plasma membrane of Madin-Darby canine kidney cells. I. Involvement of endosomes and lysosomes. *J. Cell. Biol.* 99(3):796–782, 1984.
58. Marsh, M. et al. Rapid analytical and preparative isolation of functional endosomes by free flow electrophoresis. *J. Cell. Biol.* 104(4):875–886, 1987.
59. Bartles, J. R. et al. Biogenesis of the rat hepatocyte plasma membrane in vivo: comparison of the pathways taken by apical and basolateral proteins using subcellular fractionation. *J. Cell. Biol.* 105(3):1241–1251, 1987.
60. Nath, S. et al. Separation of enzymes from Candida boidinii crude extract by continuous flow zone electrophoresis. *Electrophoresis* 11(11):937–941, 1990.
61. Hannig, K., New aspects in preparative and analytical continuous free-flow cell electrophoresis. *Electrophoresis* 3:235–243, 1982.
62. Heidrich, H. G. Free-flow particle electrophoresis in cell biology and biochemistry—an updated review. In *Electrophoresis '88*, (Schafer-Nielson, C. Ed.) VCH Publishers, Weinheim, Germany, pp. 187–200, 1988.
63. Wagner, H. and Heinrich, J. Free-Flow-Electrophoresis for the separation and purification of biopolymeres. In Modern methods in protein- and nucleic acid research, (Tschesche, H. Ed.) Walter de Gruyter Verlag, Berlin, NY. pp. 69–97, 1990.
64. Hannig, K. Eine Neuentwicklung der trägerfreien, kontinuierlichen Elektrophorese. *Hoppe Seylers Z. Physiol. Chem.* 338:211–227, 1964.
65. Fawcett, J. S., Continuous-flow isoelectric focusing and isotachophoresis. *Ann. NY Acad. Sci.* 209:112–126, 1973.
66. Just, W. W., Leon, V. J. and Werner, G. Isoelectric focusing in continuous-flow electrophoresis. I. Separation of mixed red blood cells of different species. *Anal. Biochem.* 67(2):590–601, 1975.
67. Wagner, H. and Speer, W. Kontinuierliche trägerfreie isoelektrische Fokussierung. *J. Chromatogr.* 157:259–265, 1978.
68. Kostka, G. and Hannig, K. Kontinuierliche präparative Isofocussierung mit Hilfe der trägerfreien Electrophorese. In *Elektrophoreseforum '85*, (Radola, B. J. Ed.) TU München, pp. 349–352, 1985.
69. Bier, F. F. and Egen, N. B. Large-scale recycling electrofocusing. In *Electrofocus/78*, (Westerfield, J. G. Ed.) Elsevier, New York, 1979.
70. Bier, F. F. et al. New developments in isoelectric focusing. In *Structure and Biological function*, (Meierhofer, J. Ed.). Pierce Chemical Company, Rockford, IL, USA, pp. 79–89, 1979.
71. Egen, N. B. et al. A new preparative isoelectric focusing device. In *Electrophoresis '83*, (Hirai, H. Ed.) *Walter de Gruyter, Berlin*, pp. 547–550, 1984.
72. Righetti, P. G., Wenisch, E. and Faupel, M. Preparative Protein Purification in a Multicompartment Electrolyzer with Immobiline Membranes. *J. Chromatogr.* 475:293–309, 1989.

73. Morrison, D. R. Cell electrophoresis in microgravity: past and future. In *Cell Electrophoresis* (Bauer, J. Ed.) CRC Press, Boca Raton, FL, pp. 283–313, 1994.
74. Hannig, K. and Bauer, J. Free flow electrophoresis in space shuttle program (Biotex). *Adv. Space Res.* 9(11):91–96, 1989.
75. Kobayashi, H., Ishii, N. and Nagaoka, S. Bioprocessing in microgravity: free flow electrophoresis of C. elegans DNA. *J. Biotechnol.* 47(2–3):367–376, 1996.
76. Clifton, M. J. et al. Purification of biological molecules by continuous flow electrophoresis in the Second International Microgravity Laboratory. *J. Biotechnol.* 47(2–3):341–352, 1996.
77. Hirokawa, T. et al. Free-flow isotachophoresis under micro-gravity. *Biol. Sci. Space.* 14(3):260–261, 2000.
78. Bauer, J. et al. Electrophoresis in space. *Adv. Space Biol. Med.* 7:163–212, 1999.
79. Kuellertz, G., Meyer, S. and Fischer, G. Differentiation by preparative continuous free flow-isoelectric focusing of cyclosporin A inhibitable peptidyl-prolyl cis/trans isomerase of human erythrocytes. *Electrophoresis* 15(7):960–967, 1994.
80. Bondy, B. et al. Sodium chloride in separation medium enhances cell compatibility of free flow electrophoresis. *Electrophoresis*, 16(1):92–97, 1995.
81. Anderson, N. L. and Anderson, N. G. Proteome and proteomics: new technologies, new concepts, and new words. *Electrophoresis,* 19(11):1853–1861, 1998.
82. Righetti, P. G., Castagna, A. and Herbert, B. Prefractionation techniques in proteome analysis. *Anal. Chem.* 73(11):320A–326A, 2001.
83. Poggel, M. and Melin, T. Free-flow zone electrophoresis: a novel approach and scale-up for preparative protein separation. *Electrophoresis* 22(6):1008–1015, 2001.
84. Raymond, D. E., Manz, A. and Widmer, H. M. Continuous sample pretreatment using a free-flow electrophoresis device onto a silicon chip. *Anal. Chem.* 66(18):2858–2865, 1994.
85. Raymond, D. E., Manz, A. and Widmer, H. M. Continuous separation of high molecular weight compounds using a microliter volume free-flow electrophoresis microstructure. Anal. Chem. 68(15):2515–2522, 1996.
86. Macounova, K., Cabrera, C. R. and Yager, P. Concentration and separation of proteins in microfluidic channels on the basis of transverse IEF. *Anal. Chem.* 73(7):1627–1633, 2001.
87. Macounova, K. et al. Generation of natural pH gradients in microfluidic channels for use in isoelectric focusing. *Anal. Chem.* 72(16):3745–3751, 2000.
88. Cabrera, C. R., Finlayson, B. and Yager, P. Formation of natural pH gradients in a microfluidic device under flow conditions: model and experimental validation. *Anal. Chem.* 73(3):658–666, 2001.
89. Cabrera, C. R. and Yager, P. Continuous concentration of bacteria in a microfluidic flow cell using electrokinetic techniques. *Electrophoresis* 22(2):355–362, 2001.
90. Mazereeuw, M. et al. Free flow electrophoresis device for continuous on-line separation in analytical systems. An application in biochemical detection. *Anal. Chem.* 72(16):3881–3886, 2000.
91. Chartogne, A., Tjaden, U. R. and Van der Greef, J. A free-flow electrophoresis chip device for interfacing capillary isoelectric focusing on-line with electrospray mass spectrometry. *Rapid Commun. Mass Spectrom.* 14(14):1269–1274, 2000.
92. Shinohara, E. et al. Microfabricated free flow electrophoresis module for sample preparations. *Anal. Sci.* 17(Suppl.):i441–i443, 2001.
93. Baldermann, C. et al. The regulated outer membrane protein Omp21 from Comamonas acidovorans is identified as a member of a new family of eight-stranded beta-sheet proteins by its sequence properties. *J. Bacteriol.* 180(15):3741–3749, 1998.
94. Maida, R. et al. Three pheromone-binding proteins in olfactory sensilla of the two silkmoth species Antheraea polyphemus and Antheraea pernyi. *Eur. J. Biochem.* 267(10):2899–2908, 2000.
95. Bernardo, K. et al. Purification and characterization of a magnesium-dependent neutral sphingomyelinase from bovine brain. *J. Biol. Chem.*, 275(11):7641–7647, 2000.

96. Alpert, A. J. Hydrophilic-interaction chromatography for the separation of peptides, nucleic acids and other polar compounds. *J. Chromatogr.* 499:177–196, 1990.
97. Jeno, P. et al. Desalting electroeluted proteins with hydrophilic interaction chromatography. *Anal. Biochem.* 215(2):292–298, 1993.
98. Zhang, J. and Wang, D. I. Quantitative analysis and process monitoring of site-specific glycosylation microheterogeneity in recombinant human interferon-gamma from Chinese hamster ovary cell culture by hydrophilic interaction chromatography. *J. Chromatogr. B.* 712(1–2):73–82, 1998.
99. Lindner, H. et al. Separation of acetylated core histones by hydrophilic-interaction liquid chromatography. *J. Chromatogr. A* 743(1):137–144, 1996.
100. Spanik, I. and Vigh, G. Effect of feed zone width on product purity in preparative-scale; continuous free-flow isoelectric focusing separation of enantiomers. *J. Chromatogr. A* 979(1–2):123–129, 2002.
101. Spanik, I., Lim, P. and Vigh, G. Use of full-column imaging capillary isoelectric focusing for the rapid determination of the operating conditions in the preparative-scale continuous free-flow isoelectric focusing separation of enantiomers. *J. Chromatogr. A* 960(1–2):241–246, 2002.
102. Lurin, C. et al. CLC-Nt1, a putative chloride channel protein of tobacco, co-localizes with mitochondrial membrane markers. *Biochem. J.* 348:291–295, 2000.
103. Martinec, J. et al. Subcellular localization of a high affinity binding site for D-myo-inositol 1,4,5-trisphosphate from Chenopodium rubrum. *Plant Physiol,* 124(1):475–483, 2000.
104. Kang, T. et al. Subcellular distribution and cytokine- and chemokine-regulated secretion of leukolysin/MT6-MMP/MMP-25 in neutrophils. *J. Biol. Chem.* 276(24):21960–21968 2001.
105. Kushimoto, T. et al. A model for melanosome biogenesis based on the purification and analysis of early melanosomes. *Proc. Natl. Acad. Sci. USA* 98(19):10698–10703, 2001.
106. Thery, C. et al. Proteomic analysis of dendritic cell-derived exosomes: a secreted subcellular compartment distinct from apoptotic vesicles. *J. Immunol.* 166(12):7309–7318 2001.
107. Bard, F. et al. Molecular complexes that contain both c-Cbl and c-Src associate with Golgi membranes. *Eur. J. Cell. Biol.* 81(1):26–35, 2002.
108. Ellinger, I. et al. Different temperature sensitivity of endosomes involved in transport to lysosomes and transcytosis in rat hepatocytes: analysis by free-flow electrophoresis. *Electrophoresis* 23(13):2117–2129, 2002.
109. Mohr, H. and Voelkl, A. Isolation of peroxisomal subpopulations from mouse liver by immune free-flow electrophoresis. *Electrophoresis* 23(13):2130–2137, 2002.
110. Reintanz, B. et al. AtKC1, a silent Arabidopsis potassium channel alpha -subunit modulates root hair K^+ influx. *Proc. Natl. Acad. Sci. USA* 99(6):4079–4084, 2002.
111. Zischka, H. et al. Improved proteome analysis of Saccharomyces cerevisiae mitochondria by free-flow electrophoresis. *Proteomics* 3(6):906–916, 2003.
112. Hymer, W. C. et al. Mammalian pituitary growth hormone: applications of free flow electrophoresis. *Electrophoresis* 21(2):311–317, 2000.
113. Schoenberger, J. et al. Establishment and characterization of the follicular thyroid carcinoma cell line ML-1. *J. Mol. Med.* 78:102–110, 2001.
114. Stalcup, A. M. et al. Continuous free flow electrophoresis for preparative chiral separations of piperoxan using sulfated beta-cyclodextrin. *Analyst* 125(10):1719–1724, 2001.
115. Wind, M. et al. Chiral capillary electrophoresis as predictor for separation of drug enantiomers in continuous flow zone electrophoresis. *J. Chromatogr. A* 895(1–2):51–65, 2000.
116. Schneiderman, E., Gratz, S. R. and Stalcup, A. M. Optimization of preparative electrophoretic chiral separation of ritalin enantiomers. *J. Pharm. Biomed. Anal.* 27(3–4):639–650, 2002.

117. Gratz, S. R. et al. Use of dyes to investigate migration of the chiral selector in CFFE and the impact on the chiral separations. *Anal. Chem.* 73(16):3999–4005, 2001.
118. Keuth, U. et al. Separation and characterization of humic acids and metal humates by electrophoretic methods. *Electrophoresis* 19(7):1091–1096, 1998.
119. Hirokawa, T. et al. Isotachophoretic separation behavior of rare-earth EDTA chelates and analysis of minor rare-earth elements in an iron ore by bidirectional isotachophoresis-particle-induced X-ray emission. *J. Chromatogr. A* 919(2):417–426, 2001.

11 ISOELECTRIC FOCUSING AND PROTEOMICS

MELANIE Y. WHITE AND STUART J. CORDWELL

Australian Proteome Analysis Facility, Level 4, Building F7B, Macquarie University, Sydney 2109 Australia

I. INTRODUCTION

Proteomics has many broad definitions, but the simplest may refer to scale. 'Proteomics' can be defined as the ability to conduct high-throughput biochemistry or protein chemistry on a scale comparable with that achieved by molecular biology and its high-throughput counterpart, genomics. The advent of proteomics in the mid-1990s was made possible by a number of technical advancements for separating and identifying proteins, not the least of which was the sensitivity and automation capability of various mass spectrometry (MS) technologies. However, all of the original techniques used for the separation of single unique proteins from complex mixtures relied on isoelectric focusing (IEF) as a preliminary step, especially in two-dimensional gel electrophoresis (2-DE) applications. In the 10 years since the coining of the term proteome,[1,2] IEF has become

Handbook of Isoelectric Focusing and Proteomics
D. Garfin and S. Ahuja, editors.

more than just a tool utilized in 2-DE, and is now a recognized 'lynchpin' in the proteomics process. Researchers have begun to understand the value in high-resolution prefractionation steps of examining fractions with particular qualities, such as subcellular localization, prior to either 2-DE or 2-D liquid chromatography (2-DLC) for protein and peptide separation. IEF plays a central role prior to either of these applications: for example, for the separation and purification of organelles, the enrichment of high-and low-molecular mass alkaline or hydrophobic proteins, or as one of the several prefractionating devices used to enrich proteins within a given pH range and compatible with micro-range (single pH unit) 2-D gels. This chapter will deal with the proteomics aspect of IEF, focusing on its role in the prefractionation of biological samples and as the preliminary step in 2-DE, with an emphasis on reviewing most recent developments.

A. Problems in Proteomics and How to Solve Them

Currently, two main methods exist for the separation of proteins in proteomics projects: (i) 2-DE and (ii) 2-DLC. This chapter will concentrate on 2-DE technology; however, many of the IEF-based prefractionating devices now available are suitable for use with 2-DLC applications, and furthermore, many of the inherent problems associated with 2-DE also pertain to 2-DLC—for example, an inability to solubilize highly hydrophobic proteins. Despite these recent advances in alternative technologies, including multi-dimensional chromatography[3] and isotope-coded affinity tags[4] combined with MS, 2-DE remains the method of choice for performing high-resolution separation[5] of complex protein mixtures.[5,6] This is because 2-DE is capable of simultaneously resolving between 3000 and 10,000 protein species, and is the only method suitable for visualizing and purifying protein 'isoforms,' namely those protein species containing post-translational modifications (PTM) that subtly alter the molecular mass (M_r) and isoelectric point (pI) of a predicted protein. Furthermore, 2-DE remains a preferred technology for most laboratories because of the relatively low cost of its equipment in comparison with that of other separation methods. Scientifically, several studies have shown that an approach utilizing the power of both 2-DE and 2-DLC results in the best possible proteome coverage.[7] In recent times, the resolving power of 2-DE has been further enhanced by improved methodology for biological sample preparation as well as incremental improvements in 2-DE technology, including micro-range (1.0–1.5 pH units) immobilized pH gradients (IPG),[8–10] better fluorescent dyes,[11] and more reproducible precast second dimension sodium dodecyl sulfate–polyacrylamide gel electrophoresis (SDS–PAGE) gels.

Despite these improvements, several drawbacks with 2-DE still remain. We are only seeing the "*tip of the proteome iceberg*"[12] due to the underrepresentation of: (i) hydrophobic or membrane-associated proteins; (ii) highly alkaline proteins; (iii) high- and low-molecular-mass proteins; and (iv) lower-abundance proteins due to the relative insensitivity of available stains and the dominance of high-abundance "housekeeping" proteins. This is especially apparent in samples where one or two individual protein species account for a significant fraction of the total protein present (e.g., albumin in human plasma). One successful method for overcoming these limitations has been achieved by taking a subproteomics approach.[8] This can be performed by prefractionation to purify a subcellular compartment or organelle,[13,14] or by using the physical, chemical, or functional properties of a protein or class of proteins to enrich a particular subset prior to 2-DE or 2-DLC. Such an approach serves two purposes: (i) the group of proteins associated with a given subcellular fraction or organelle can be studied with relative specificity and (ii) the overall complexity of the 2-DE pattern is reduced when compared with whole-cell preparations, making it possible to visualize more lower abundance proteins.

II. THE PROTEOMICS WORKFLOW

Proteomics is generally applied to questions of biological significance, and hence the experimental design is aimed at finding biomarkers, irrespective of whether they be they diagnostic of disease, quality in agricultural crops, or pathogenicity factors in infectious disease. The aim is to determine, among a complex variety of unique proteins, those that are associated with performing a particular biological process. Therefore, the goal of any proteomics researcher must be to maximize the amount of the functional proteome that can be viewed at any given point in time, or to prefractionate samples such that an increasingly specific subset of the functional proteome most relevant to the biological process under investigation is purified. The proteomics "workflow" (Figure 1) therefore begins with the prefractionation of complex samples based on physical (e.g., pI or M_r), chemical (e.g., PTM), or functional (e.g., ligand- or protein-binding) properties, or upon subcellular localization (e.g., membrane, extracellular, nuclear, and mitochondrial proteins).

These methods use selective, sequential solubilization[15] or the generation of relatively pure subcellular fractions as well as an increasing array of chromatographic approaches for the enrichment of proteins with a selectable property.[16–18] Essentially, sequential solubilization fractionates whole cell or tissue by removing highly hydrophilic proteins generally associated with the cytosol before attempting to extract proteins from previously insoluble material, using progressively more "aggressive" solubilization buffers (Figure 2). The resulting 2-DE patterns from these experiments share a degree of similarity, yet also show the enrichment of

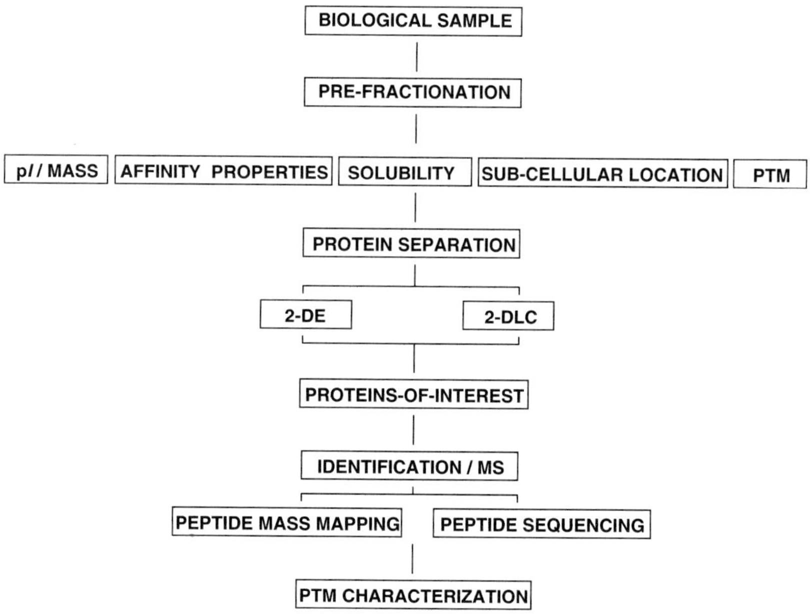

FIGURE 1 **Schematic representation of a 'typical' proteomics experiment. Biological samples are prefractionated, separated using 2-DE or 2-DLC, subjected to image or informatics analysis to determine "proteins-of-interest", which are then identified and further characterized by MS.**

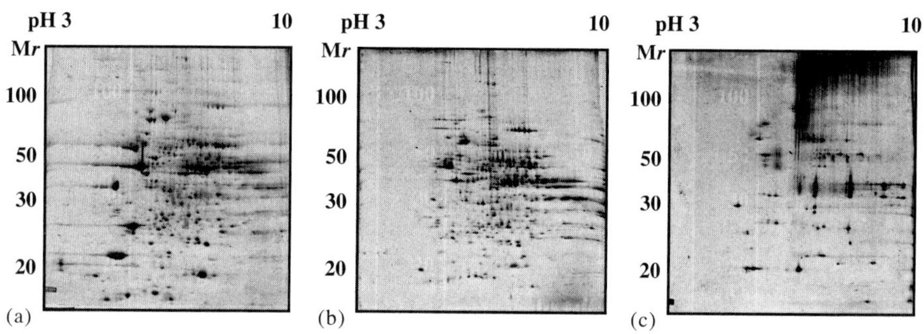

FIGURE 2 **Subcellular fractionation of proteins from rabbit myocardium. Comparison of (a) whole tissue proteins, (b) hydrophilic or cytosolic proteins and (c) purified membrane proteins.**

hydrophobic proteins in the later fractions. Clearly, this is dependent on the sample type, and as yet, no group has conducted a large-scale identification project to determine the degree of protein "carry-over" between fractions, i.e., what level of enrichment can actually be achieved for a given sample type.

Organellar fractionation prior to sample solubilization has been utilized for membrane-specific proteins, as well as nuclear, mitochondrial, and several other sub-cellular fractions, using a variety of biochemical extraction techniques prior to 2-DE or 2-DLC. The most important element in these types of approaches is that the subcellular fraction must be close to true purity or abundant contaminants from other fractions will interfere. Subcellular proteomics has been performed on fractions containing the nucleus,[19] mitochondria,[20–22] and prokaryotic and eukaryotic membranes.[23–26] Furthermore, a range of commercial kits are now available that standardize many of these methods. Such extractions have different degrees of success depending on the sample type. For example, proteins from the outer membrane of Gram-negative bacteria can be isolated almost to purity; proteins from mammalian plasma membranes, however, will more likely be significantly enriched, but not to absolute purity. Again, no studies have been performed to ascertain the level of enrichment for more than a single sample type.

Although not strictly relevant to the current chapter, the chromatographic approach to sample prefractionation has a vast degree of potential for proteomics applications and is reviewed in Righetti et al.[18] and Lee and Lee.[27] Affinity chromatography has several applications including: (i) the removal of abundant proteins, especially serum albumin from plasma[28,29] or cerebrospinal fluid,[30] which reduce the separating capacity of 2-DE; (ii) the concentration of protein samples; (iii) the selective binding of proteins with a selectable property (e.g., phosphoproteins and metal-binding proteins using immobilized metal-affinity chromatography, IMAC,[31] or various ligand-binding proteins such as those that bind ATP[32] or calcium[33]); and (iv) the study of protein complexes. Many of these approaches have applications in "peptidomics," or in the selective purification of a particular modified peptide prior to mass spectrometric analysis.

Whole cell or tissue extracts, or prefractionated protein samples, are then separated by SDS–PAGE, 2-DE, or 2-DLC prior to the identification of the purified components by MS. The choice of separation technology greatly depends on the complexity of the sample. While whole protein mixtures can be separated using 2-DE, SDS–PAGE may be suitable for purified protein complexes, and 2-DLC of complex mixtures is utilized following proteolytic digestion to create a peptide mix that can be separated by LC. Peptides derived from isolated gel spots and bands, or purified via 2-DLC, are then used to identify the proteins by a variety of MS techniques (reviewed in Reference 34). The final step in the traditional proteomics process is to characterize the nature of protein PTMs. While this is not the focus of the current review, IEF-based 2-DE is currently the best method for viewing modified proteins. A variety of novel staining techniques, specific for phosphoproteins[35,36] and glycoproteins,[37] are now available, making the determination of PTM-containing proteins more efficient.

III. IEF FOR PREFRACTIONATION

IEF is making a significant contribution to prefractionating proteins prior to conventional separation (2-DE and 2-DLC approaches), and to protein identification and characterization (mainly performed using MS). There are two main approaches where IEF has made an impact on prefractionation: (i) the separation of subcellular fractions and organelles and (ii) the separation of proteins into pI fractions suitable for use with micro-range IPG or 2-DLC-MS, thus aiding in the identification of lower abundance proteins. Each of these approaches are leading to a more improved resolution and proteome "coverage" than that can be achieved with currently available conventional technology.

A. Prefractionation of Organelles

As mentioned previously, subcellular fractionation is an important means of improving the information that can be acquired using proteomics technologies. Subcellular fractionation is often performed using density-gradient centrifugation, differential extraction, or other biochemical methods. In recent times, these methods have been supplemented by using IEF to further enrich proteins from a particular organelle. Zischka et al.[38] utilized a free-flow electrophoresis (FFE) device to purify further mitochondria from *Saccharomyces cerevisiae* following density-gradient ultracentrifugation. In this particular study, only 2% of the identified proteins were known to be non-mitochondrial in comparison with the 16% of non-mitochondrial proteins identified when only the centrifugation step was performed. Furthermore, 43 mitochondrial proteins were identified only in the centrifugation-FFE fractions. This method clearly has substantial potential for providing very pure organellar fractions prior to 2-DE.

B. Prefractionating Devices

Several commercial prefractionating devices now exist for the separation of complex protein mixtures prior to 2-DE or 2-DLC and subsequent to MS identification and characterization of the individual purified components. The methods based on IEF include the Gradiflow (Gradipore), Rotofor (Bio-Rad), FFE (Tecan), and multicompartment electrolyzers (Proteome Systems, Invitrogen). Rotofor is a liquid-phase IEF-based prefractionation device that separates proteins in free solution using the buffering capability of carrier ampholytes (CA) to create a linear pH gradient. Proteins are fractionated into sample chambers (typically 20) separated by permeable screens, the resulting fractions are collected, and CA are removed, prior to 2-DE or 2-DLC.[39] This approach has been used to separate further proteins from cerebrospinal fluid[40,41] and ovarian carcinoma cells.[42] Recent work has modified the Rotofor to perform liquid-phase IEF in the absence of CA by employing polyacrylamide membranes with embedded Immobilines.[43]

Gradiflow is a preparative electrophoresis apparatus that incorporates a membrane-based separation using thin polyacrylamide membranes with variable pore sizes to allow both size- and charge-based separations.[44] Essentially two fractions can be achieved, and then further rounds of purification are possible on these fractions. As such it is a simple device for removing abundant proteins including serum albumin,[45] prefractionating complex mixtures prior to micro-range 2-DE,[46,47] and for the analysis of proteins incompatible with 2-DE, for example, basic proteins.[48] For compatibility with micro-range IPG 2-DE, Gradiflow may not be ideal as only 2 pI fractions can be achieved in a single run. Here multicompartment devices[49–51] may be more suitable. These devices are based on isoelectric membranes using immobiline technology and have several distinct advantages, including their downstream compatibility with IPG-2-DE without further sample cleanup or removal of CA, reduced sample precipitation, and the flexibility to create membranes specific for the removal of a given protein or protein fraction. These advantages are well documented in Reference 18. The MCE has recently been applied to the separation of yeast cell lysates[12] with excellent results, including the separation and identification of membrane-associated and lower abundance proteins. Similar technology utilizing thin immobiline-containing polyacrylamide membranes has been developed by Zuo and Speicher[52,53] and has been used to detect low abundance and other "difficult" proteins in human breast cancer.[54] This device has been used to perform prefractionation based on pI prior to micro-range IPG 2-DE with compatible fractions for minimal sample "cross-over" between IPGs, hence providing improved quantitation.[55] These authors have also been capable of separating prefractionated proteins using 0.5 pH unit fractions in conjunction with single pH unit IPGs, although there is obviously an opportunity to use even narrower IPG pH ranges, especially in those areas where substantial numbers of expressed proteins are usually found (e.g., pH range between 4.5 and 5.5). This principle is also found in Görg et al.,[56] where Sephadex IEF is performed as a "third" dimension prior to IPG-2-DE. The advantage of this system is its simplicity, as the proteins are treated with sample buffers identical to those used in IPG 2-DE, and the resulting Sephadex fractions are simply cut with a scalpel and applied to the IPG strip.

Unfortunately, for proteome-wide coverage using 2-DE, increased numbers of fractions (including isoelectric and subcellular fractionation or a combination of both) mean that a potentially vast number of gels must be run to provide real coverage rather than just a proteome "snapshot." This fact combined with the substantial and necessary number of replicates that must be run to provide meaningful statistical analyses means that such an approach is only viable for very few laboratories. Finally, a concerted attempt to use this type of combined approach to generate significant data has yet to be shown as viable for any biological problem.

IV. IEF IN TWO-DIMENSIONAL ELECTROPHORESIS

IEF of proteins is the technology that has allowed two-dimensional gel electrophoresis to become the key methodology for protein separation and the purification of individual components from complex mixtures.[57] Two methods exist for this process: first, the separation of proteins in capillary rod-gels where the acrylamide gel matrix contains CA molecules which are "prefocused" to create the pH gradient (reviewed in Reference 58); and secondly IPG technology that utilizes well-characterized acrylamido buffers to create a stable pH gradient in a thin acrylamide strip (reviewed in Reference 5). The CA method suffers from several drawbacks, including an inability to load preparative amounts of sample, reproducibility problems associated with batch-to-batch variations of CA preparations, difficulties in pouring and handling the resulting IEF gels, and cathodic drift caused by the destabilization of the pH gradient at alkaline pH. Despite these limitations, several outstanding research groups have generated a wealth of proteomic data using CA-dependent IEF in the first dimension of 2-DE including those of Celis,[59–61] VanBogelen,[62,63] and Hecker.[64,65]

IPG technology revolutionized 2-DE and was an important factor, along with the advent of sensitive, high-throughput MALDI–TOF and ESI–MS, behind the proliferation of proteomics studies.[5] IPG IEF in the first dimension of 2-DE has allowed for better reproducibility both run-to-run and between laboratories following standard procedures with commercially available reagents and precast IPG strips,[66,67] higher loading capacity and in-gel rehydration of protein samples, and improved ease of use. IPG strips are also commercially available from a number of vendors in a wide variety of pH gradients ranging from very wide (pH 4–12 and 3–10), medium (pH 3–6, 4–7, 5–8, 7–10, 6–9, 6–11, etc.) to micro-range (single pH unit), and in a variety of strip lengths suitable for either high resolution (17, 18, 24, and 30 cm) or rapid analysis (7 and 11 cm) proteomics applications.

As mentioned previously, despite these advances, several problems remain with 2-DE technology for high-resolution proteomics applications. The final sections of this chapter deal with how modifications to IEF-based prefractionation and IPG IEF can be applied to overcome some of these limitations.

A. Alkaline Proteins

One of the main problems traditionally associated with 2-DE is its inability to reproducibly array highly alkaline proteins (pI > 9). These can be lost due to cathodic drift during CA IEF, or the lack of suitable buffering molecules at highly alkaline pH.[68] Prior to the advent of IPG technology, non-equilibrium pH gradient electrophoresis (NEPHGE) was utilized to provide the separation of basic proteins.[69] IPG IEF is possible in the basic pH range, even up to a pI of 12.0.[5,68,70] The separation of alkaline proteins is challenging for several reasons: (i) active water transport toward the anode (reverse electroendosmosis) caused by the strong positive charge of basic acrylamido buffers; (ii) the hydrolysis of acrylamide to acrylic acid at alkaline pH; and (iii) the migration of reducing agents, mainly dithiothreitol (DTT), leading to reduced gel quality ("streaky" 2-DE patterns) and diminished reproducibility. In an attempt to overcome these limitations, some special precautions must be taken to ensure quality separations: (i) proteins must be cup-loaded at the anode, especially where prior isoelectric prefractionation has not occurred; (ii) samples may be treated with hydroxyethyl disulfide allowing oxidation of thiol groups in disulfide containing proteins;[71] (iii) dimethylacrylamide (DMA[68]) or *N*-acryloylaminoethoxyethanol (AAEE[72]) may be used instead of acrylamide since these matrices have been shown to resist alkaline hydrolysis at basic pH; (iv) non-ionic reducing agents such as tributylphosphine (TBP) can be used to replace DTT,[73] thus minimizing the transport of reducing agent out of the IPG during IEF; and (v) isopropanol or glycerol can be added to sample buffers to reduce the electroendosmotic effects.[73] Despite these improvements, high-quality protein separations at very alkaline pH values using 2-DE remain rare.

In one recent study, a concerted attempt was made to characterize highly alkaline proteins in the bacterium *Helicobacter pylori*.[48] This study utilized the Gradiflow prefractionation device to examine 2-DE-incompatible proteins and compare them with separations using pH 6–11 and 9–12 IPGs. *H. pylori* is an excellent model for this type of work since it contains a higher proportion of alkaline proteins than acidic ones (Figure 3). Prefractionation allowed a collection of proteins with pI > 9.0 to be collected and analysed using SDS–PAGE and LC-MS/MS to reveal novel protein species that could not be detected using 2-DE. Another group determined that subcellular fractions taken from rat liver were enriched for proteins with different pI properties as well (e.g., mitochondrial extracts were enriched for basic proteins).[74] While this study was performed with 2-DLC/MS-MS, very few groups have combined isoelectric prefractionation with this separation and identification strategy.

B. Low Abundance Proteins and Micro-range "Zoom" IPG 2-DE

2-DE currently fails to resolve many lower abundance proteins for two major reasons: (i) the "dynamic range" of protein abundances in most cells and tissues is too wide to accommodate a good focusing of

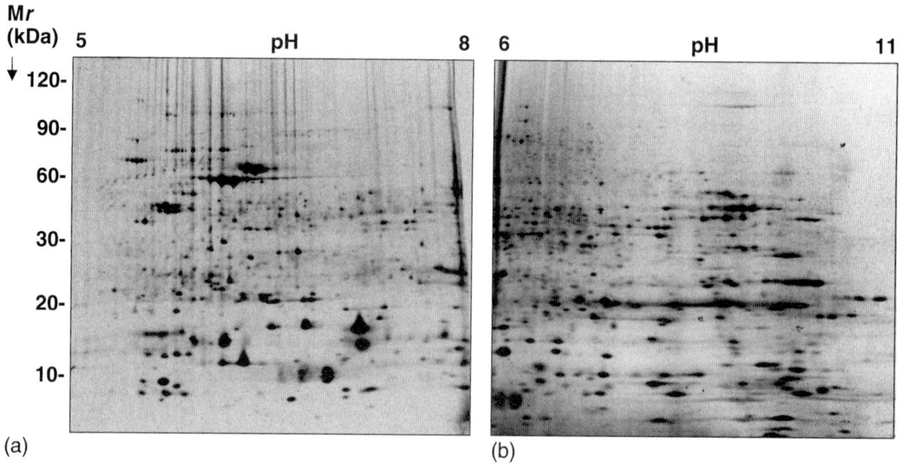

FIGURE 3 **2-DE gels using pH 5–8 (a) and pH 6–11 (b) IPG IEF separation of *H. pylori* whole cell proteins.**

abundant proteins while resolving lower abundance proteins and (ii) the limit of detection of currently available fluorescent and visible stains. One method used to overcome these limitations is to utilize affinity or other methods for the removal of abundant proteins. Such methods have been mainly applied to human serum for the removal of serum albumin, immunoglobulins, serotransferrin, etc., thus enabling the detection of lower abundance biomarkers.[28,29] However, many of these methods result in the removal of non-targeted proteins and hence reduce the quantity of lower abundance or bound proteins as well. Isoelectric prefractionating devices have also been used for abundant protein removal, by restricting the pI of the separation to remove only a single major constituent.[45]

For applications using 2-DE, many groups have begun to utilize micro-range, or "zoom" IPG strips (1 pH unit), either alone or in conjunction with compatibly prefractionated samples. The use of micro-range IPGs serves two purposes: first, to increase the separating area versus pH range ratio of the IEF dimension, and second, to allow a higher concentration of protein sample to be applied to the IPG (Figure 4). This results in the ability to visualize more low abundance proteins. Furthermore, the effect is further increased when isoelectrically prefractionated samples are applied to the micro-range IPG. This is a simple loading effect, for example, 1 mg of complex mixture applied to a micro-range gel in the range of pH 4–5 may result in the separation of effectively 500–750 ng of proteins with pI between 4 and 5. Where IEF prefractionation is performed, several mg of complex protein mixture might be prefractionated such that exactly 1 mg of protein with pI between 4 and 5 is applied to the compatible IPG strip. This also results in less sample buildup at the IPG extremes that might result in less efficient focusing. Several studies have examined the utility of micro-range

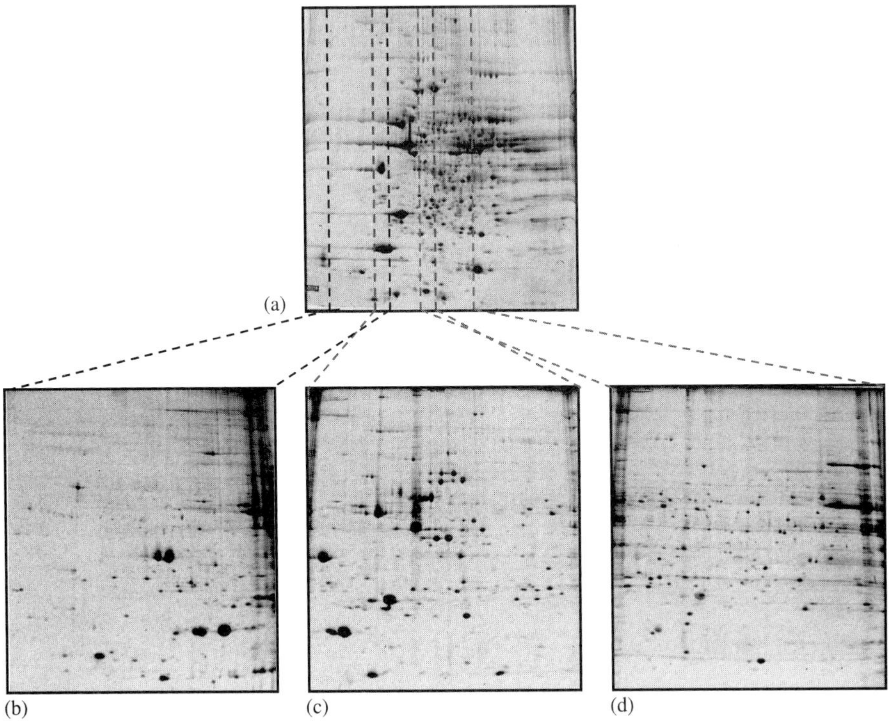

FIGURE 4 **2-DE gels utilizing micro-range (1.0 pH unit) IPG IEF. Rabbit myocardium whole tissue proteins separated on (a) pH 3–10; (b) pH 3.9–5.1; (c) pH 4.7–5.9; and (d) pH 5.5–6.7 IPGs.**

IPG 2-DE both alone and in conjunction with other prefractionation techniques including subcellular fractionation or differential solubility, and isoelectric fractionation as described above.[5,8,10,55]

One disadvantage of the creation of "composite" 2-DE preparations comprised of several parallel, overlapping narrow-range IPG 2-DE gels is that a substantial amount of work is required to produce 4 or 5 gels per sample, rather than only 1 or 2. This combined with the use of subcellular prefractionation, and the required number of replicates necessary for meaningful statistical analyses, which mean that only a handful of laboratories are capable of conducting the research. The most useful applications for these IPG 2-DE gels appears to be in utilizing micro-ranges to provide the requisite separation of protein isoforms for mass spectrometric characterization of PTM (Figure 5).[75]

C. High and Low Molecular Mass Proteins

Another significant problem associated with 2-DE is the underrepresentation of high- and low-molecular-mass proteins and peptides. These problems occur due to the relative acrylamide pore size separation of

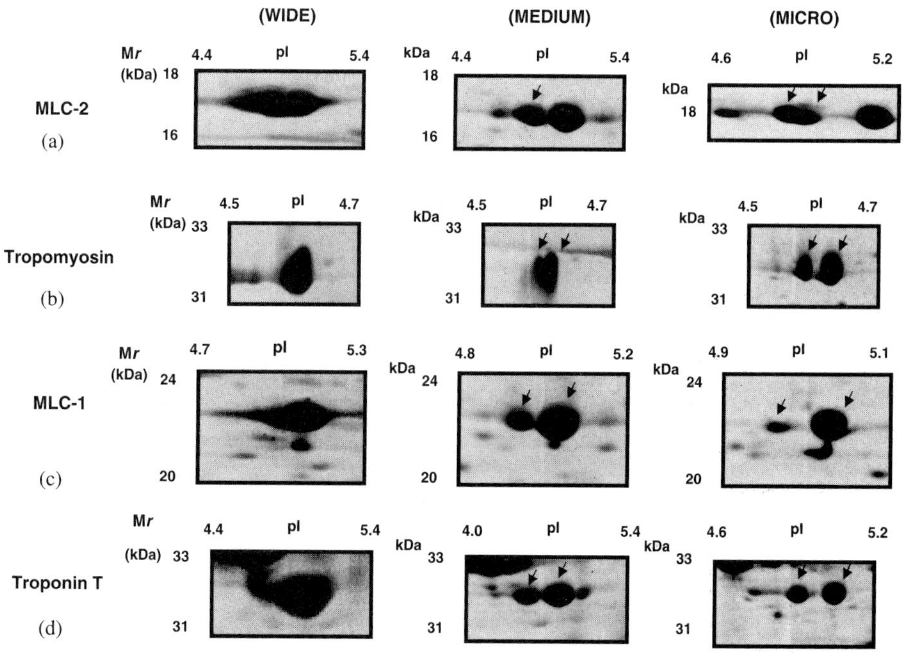

FIGURE 5 2-DE gels of separated proteins from rabbit myocardium showing enhanced resolution of protein isoforms utilizing micro-range (1.0 pH unit) IPGs: (a) myosin light chain 2, (b) tropomyosin; (c) myosin light chain 1, and (d) troponin T.

both the first-dimension IPG strip, and the %*T* (monomer) gradient used in the SDS–PAGE dimension. The IPG strip pore size is 4%*T* in commercially available strips, which excludes higher mass proteins >120–150 kDa. Recent work has attempted to improve the recovery of these proteins using lower %*T* IPG strips in the IEF dimension, to as low as 3%*T*.[76,77] This has a two-fold effect: to improve the total amount of protein that enters the strip in the first dimension (either following cup-loading or in-gel rehydration), and also to increase the proportion of larger proteins on the ensuing 2-DE gels. This second effect is only possible if compatible %*T* SDS–PAGE gels are incorporated in the second dimension. Another group has utilized agarose gels in the IEF dimension to improve the separation of high–molecular mass proteins.[78] Lower molecular mass proteins (i.e., <10 kDa) are typically resolved using a Tris-tricine SDS–PAGE system.[79] This simple modification of the 2-DE procedure has allowed for varying studies of low mass proteins and peptides including those from amyloid brain plaques,[80] *Haemophilus influenzae*[81] and *Escherichia coli.*[82]

Prefractionation, using both isoelectric and size-exclusion type methods, has also been successfully utilized for the analysis of both high- and low-mass proteins. Gradiflow uses both size- and charge-based fractionation, i.e., fractions with a molecular mass cutoff of >100 or 200 kDa, or <5 kDa can be achieved easily. The proteins within the fractions can then be precipitated and analysed by either SDS–PAGE or 2-DLC MS, because the fractions are incompatible with 2-DE. Another preparative electrophoresis device is the PrepCell (Bio-Rad), which allows for the collection of molecular mass fractions compatible with SDS–PAGE, 2-DE or 2-DLC separations. This has been used to characterize high-and low-abundance proteins from rat liver and brain cytosol.[83,84]

V. CONCLUSIONS

IEF remains an integral part of proteome analysis. It is central to the most commonly performed protein separation technique utilized in proteomics (2-DE), and in recent times has become essential for protein prefractionation prior to 2-DE, or other, separation techniques. As such it is a crucial element in allowing proteomics to access more than just the most abundant and readily solubilized proteins, and hence is a pre-requisite for the proteomic deep drilling of even the simplest of organisms.

GLOSSARY

2-DE Two-dimensional gel electrophoresis.
2-DLC Two-dimensional liquid chromatography.
CA Carrier ampholytes.
FFE Free-flow electrophoresis.
IEF Isoelectric focusing.
IPG Immobilized pH gradient.
MCE Multi-compartment electrolyzer.
MS Mass spectrometry.
NEPHGE Non-equilibrium pH gradient electrophoresis.
PTM Post-translational modification.
SDS–PAGE Sodium dodecyl sulfate–polyacrylamide gel electrophoresis.

ACKNOWLEDGMENTS

This work was facilitated by access to the Australian Proteome Analysis Facility (APAF) funded under the Australian government Major National Research Facility (MNRF) program. Bio-Rad Laboratories and the National Heart Foundation of Australia are thanked for funding. MYW is the recipient of an APAF-Bio-Rad Postgraduate Award.

Column

pH of elution buffer entering the column		pH of mobile phase segments / pH of stat. phase segments			pH of mobile phase exiting the column
pH of column equilibrated with application buffer at the step change to the elution buffer	5	8	8	8	8
		8	8	8	
Shift of mobile phase	5	5	8	8	8
		8	8	8	
Equilibrium pH values after shift	5	7	8	8	8
		7	8	8	
Shift of mobile phase	5	5	7	8	8
		7	8	8	
Equilibrium pH values after shift	5	6.33	7.67	8	8
		6.33	7.67	8	
Shift of mobile phase	5	5	6.33	7.67	8
		6.33	7.67	8	
Equilibrium pH values after shift	5	5.89	7.22	7.89	8
		5.89	7.22	7.89	
Shift of mobile phase	5	5	5.89	7.22	7.89
		5.89	7.22	7.89	
Equilibrium pH values after shift	5	5.59	6.78	7.67	7.89
		5.59	6.78	7.67	
Shift of mobile phase	5	5	5.59	6.78	7.67
		5.59	6.78	7.67	
Equilibrium pH values after shift	5	5.39	6.38	7.37	7.67
		5.39	6.38	7.37	

FIGURE 2 Example calculations of pH in a weak anion-exchange column used in a conventional chromatofocusing setup employing the buffer interaction model. The chromatofocusing column is divided into three segments. The pH of the stationary (bottom) and mobile (top) phases is indicated immediately after a shift of mobile phase to the adjacent downstream segment and after equilibrium is reached (calculated with Equation (1)) prior to the next shift. Application and elution buffers have pH values of 8 and 5, respectively, and $\beta_s = 2\beta_m$.

Transfer (Shift) Number	Seg 1	Seg 2	Seg 3	Seg 4	Seg 5	Seg 6	Seg 7	Seg 8	Seg 9	Seg 10
0	**8.00**	8.00	8.00	8.00	8.00	8.00	8.00	8.00	8.00	8.00
1	**7.00**	**8.00**	8.00	8.00	8.00	8.00	8.00	8.00	8.00	8.00
2	**6.33**	**7.67**	8.00	8.00	8.00	8.00	8.00	8.00	8.00	8.00
3	**5.89**	7.22	**7.89**	8.00	8.00	8.00	8.00	8.00	8.00	8.00
4	**5.59**	6.78	**7.67**	7.96	8.00	8.00	8.00	8.00	8.00	8.00
5	**5.40**	6.38	7.37	**7.86**	7.99	8.00	8.00	8.00	8.00	8.00
6	**5.26**	6.05	7.04	**7.70**	7.95	8.00	8.00	8.00	8.00	8.00
7	**5.18**	5.79	6.71	7.48	**7.86**	7.98	8.00	8.00	8.00	8.00
8	**5.12**	5.59	6.40	7.22	**7.74**	7.94	7.99	8.00	8.00	8.00
9	5.08	**5.43**	6.13	6.95	**7.57**	7.87	7.98	8.00	8.00	8.00
10	5.05	**5.31**	5.90	6.68	7.36	**7.77**	7.94	7.99	8.00	8.00
11	5.03	**5.23**	5.70	6.42	7.13	**7.63**	7.88	7.97	8.00	8.00
12	5.02	**5.16**	5.54	6.18	6.89	7.47	**7.80**	7.94	7.99	8.00
13	5.02	**5.12**	5.42	5.97	6.66	7.28	**7.69**	7.90	7.97	8.00
14	5.01	5.08	**5.32**	5.78	6.43	7.07	**7.55**	7.83	7.95	7.99
15	5.01	5.06	**5.24**	5.63	6.21	6.86	7.39	**7.74**	7.91	7.97
16	5.00	5.04	**5.18**	5.50	6.02	6.64	7.21	**7.62**	7.85	7.95
17	5.00	5.03	**5.13**	5.39	5.84	6.43	7.02	7.48	**7.77**	7.92
18	5.00	5.02	5.10	**5.30**	5.69	6.24	6.83	7.33	**7.68**	7.87
19	5.00	5.01	5.07	**5.24**	5.56	6.06	6.63	7.16	**7.56**	7.81
20	5.00	5.01	5.05	**5.18**	5.45	5.89	6.44	6.98	7.43	**7.72**
21	5.00	5.01	5.04	**5.14**	5.36	5.75	6.26	6.80	7.28	**7.63**
22	5.00	5.00	5.03	**5.11**	5.29	5.62	6.09	6.62	7.12	**7.51**
23	5.00	5.00	5.02	5.08	**5.23**	5.51	5.93	6.44	6.95	7.38
24	5.00	5.00	5.01	5.06	**5.18**	5.41	5.79	6.27	6.78	7.24
25	5.00	5.00	5.01	5.04	**5.14**	5.34	5.66	6.11	6.61	7.09
26	5.00	5.00	5.01	5.03	**5.11**	5.27	5.56	5.96	6.45	6.93
27	5.00	5.00	5.01	5.02	5.08	**5.22**	5.46	5.83	6.28	6.77
28	5.00	5.00	5.00	5.02	5.06	**5.17**	5.38	5.70	6.13	6.61
29	5.00	5.00	5.00	5.01	5.05	**5.14**	5.31	5.60	5.99	6.45
30	5.00	5.00	5.00	5.01	5.04	**5.11**	5.25	5.50	5.86	6.30
31	5.00	5.00	5.00	5.01	5.03	5.08	**5.20**	5.42	5.74	6.15
32	5.00	5.00	5.00	5.01	5.02	5.06	**5.16**	5.35	5.63	6.01
33	5.00	5.00	5.00	5.00	5.02	5.05	**5.13**	5.29	5.54	5.89
34	5.00	5.00	5.00	5.00	5.01	5.04	5.10	**5.23**	5.45	5.77
35	5.00	5.00	5.00	5.00	5.01	5.03	5.08	**5.19**	5.38	5.66
36	5.00	5.00	5.00	5.00	5.01	5.02	5.06	**5.15**	5.32	5.57
37	5.00	5.00	5.00	5.00	5.00	5.02	5.05	**5.12**	5.26	5.48
38	5.00	5.00	5.00	5.00	5.00	5.01	5.04	5.10	**5.22**	5.41
39	5.00	5.00	5.00	5.00	5.00	5.01	5.03	5.08	**5.18**	5.35
40	5.00	5.00	5.00	5.00	5.00	5.01	5.02	5.06	**5.14**	5.29
41	5.00	5.00	5.00	5.00	5.00	5.01	5.02	5.05	**5.12**	5.24
42	5.00	5.00	5.00	5.00	5.00	5.00	5.01	5.04	5.10	**5.20**
43	5.00	5.00	5.00	5.00	5.00	5.00	5.01	5.03	5.08	**5.17**
44	5.00	5.00	5.00	5.00	5.00	5.00	5.01	5.02	5.06	**5.14**
45	5.00	5.00	5.00	5.00	5.00	5.00	5.01	5.02	5.05	**5.11**
46	5.00	5.00	5.00	5.00	5.00	5.00	5.00	5.01	5.04	5.09
47	5.00	5.00	5.00	5.00	5.00	5.00	5.00	5.01	5.03	5.07
48	5.00	5.00	5.00	5.00	5.00	5.00	5.00	5.01	5.02	5.06
49	5.00	5.00	5.00	5.00	5.00	5.00	5.00	5.01	5.02	5.05
50	5.00	5.00	5.00	5.00	5.00	5.00	5.00	5.01	5.02	5.04
51	5.00	5.00	5.00	5.00	5.00	5.00	5.00	5.00	5.01	5.03
52	5.00	5.00	5.00	5.00	5.00	5.00	5.00	5.00	5.01	5.02
53	5.00	5.00	5.00	5.00	5.00	5.00	5.00	5.00	5.01	5.02
54	5.00	5.00	5.00	5.00	5.00	5.00	5.00	5.00	5.01	5.02
55	5.00	5.00	5.00	5.00	5.00	5.00	5.00	5.00	5.00	5.01
56	5.00	5.00	5.00	5.00	5.00	5.00	5.00	5.00	5.00	5.01
57	5.00	5.00	5.00	5.00	5.00	5.00	5.00	5.00	5.00	5.01
58	5.00	5.00	5.00	5.00	5.00	5.00	5.00	5.00	5.00	5.01
59	5.00	5.00	5.00	5.00	5.00	5.00	5.00	5.00	5.00	5.00

FIGURE 3 Equilibrium pH of weak anion-exchange column segments (both mobile and stationary phases) versus the number of mobile-phase shifts calculated with Equation 1. The chromatofocusing column is divided into 10 segments. Bold numbers indicate the positions of theoretical proteins having pI values of 7.5 (top arrowed line) and 5.1 (bottom arrowed line) versus the mobile-phase shift number. Application and elution buffers have pH values of 8 and 5, respectively, and $\beta_s = 2\beta_m$.

(downstream to pH = pI). When this protein is downstream to the pH = pI location, it becomes negatively charged and binds to the anion exchanger and thus ceases its movement in the column. The protein will remain bound to the column until the pH = pI zone passes, causing it to have a net zero charge and be released into the mobile phase. It will then proceed just past the pH = pI zone, where it develops a net negative charge and is subsequently retained on the column; and so forth down through the column. The movement of proteins through successive pH zones occurs because the mobile phase velocity exceeds the velocity of the pH zones. This process by which a protein is continually relocated into its pH = pI zone is also the source the focusing effect in chromatofocusing, leading to narrow peak widths.

There is thus an initial separation of the proteins in distance on the column with the establishment of the internal pH gradient. This internal pH gradient however is not stagnant. It moves down the column with time as more elution buffer is pumped into the column, as seen in Figure 3. Eventually each pH will exit the column along with proteins that have a pI at that pH.

Each pH band moves through the column at a constant velocity that is different for each particular pH value. This is seen in Figure 3 for examples of two theoretical proteins having pI values of 5.1 and 7.5. The position where each protein is retained on the column (at the first column segment where the protein encounters a pH > pI) versus the number of mobile phase transfers is given in bold. It is seen that the positions at various times fit a straight line (indicated by the arrowed lines), meaning that the proteins (and their associated pH zones) travel at a constant velocity during the entire chromatofocusing process. The slopes are different for the two proteins, indicating different velocities for each different pI (or pH band); the lower the pI of a protein, the lower its velocity in the chromatofocusing column. The velocity of a particular pH band thus steadily decreases for pH bands with successively lower values. This is because the higher pH values are titrated with the greatest pH change (exposed to the front of the step gradient) and thus the change in pH for the higher pH values occurs rapidly. The lower value pH bands however have less difference between themselves and the elution buffer pH, thus the change in pH is less rapid. Also, the velocity of a particular pH band is inversely related to the pH of the elution buffer. Finally, the velocity of a pH band is also dependent on the ratio β_m/β_s. The higher the ratio, the higher the velocity of a pH band, because an aliquot of mobile-phase buffer will titrate the stationary phase to a greater extent.

Sluyterman's and Elgersma's theory of conventional chromatofocusing views pH gradients from the perspective of the internal pH gradient (dpH/dx, where x is units of column distance). This viewpoint is analogous to IEF in that pH values are assigned to a particular location on the gel. However, unlike in IEF, there is no set position for a particular pH in the column, as it is continually moving through the column. Although direct characterization of the internal gradient is not conveniently done, it can be indirectly characterized by measuring the pH at the column outlet. The outlet pH gradient correlates positively with the internal pH gradient, meaning a change in the internal pH gradient produces a proportional change in the outlet pH gradient. From an experimental viewpoint, the characterization of the outlet pH gradient (time gradient) is more practical than the characterization of the internal pH gradient (distance gradient) because separations in chromatofocusing are monitored in time, not in distance.

Higher resolution is obtained with decreased slope of the internal pH gradient (and outlet pH gradient).[1,19] Shallow slopes are obtained by decreasing the pH gradient range (i.e., using application and elution buffers with small differences in pH), increasing the column length, decreasing the concentration of the elution buffer (i.e., decreased buffer capacity of the elution buffer), or by using a different anion-exchange column with higher buffer capacity.[2,19]

2. Displacement (Frontal) Model in Anion Chromatofocusing

The displacement mechanism of chromatofocusing was first proposed for ampholyte-displacement chromatography,[27] which is a technique closely related to, if not the same as, chromatofocusing. This mechanism was mentioned by Sluyterman and Elgersma,[1] proposed by Hearn and Lyttle,[1,19] and further developed by Murel et al.[28,29] in application to chromatofocusing. The displacement mechanism proposes the establishment of an internal pH gradient within a column via a gradient distribution of different buffer components within the ion-exchange column. This internal pH gradient is established because the stronger acid components in the elution buffer will preferentially bind to the initial anion-exchange sites at the inlet, with the next strongest acid binding to the next available sites down the column and so forth, assuming that the concentrations of the buffers are low enough such that all the sites are not saturated with the strongest acid components. This retention of successively lower affinity (decreasing acidity) buffer components down the length of the column leads to an increasing pH gradient in distance from the column inlet to outlet. Note that binding only occurs for the conjugate basic (anionic) forms of these acids, which are the forms that are present in the column equilibrated to high pH.

As more elution buffer enters the column, additional stronger acid components are introduced onto the column, which displace weaker acid components immediately downstream to it, and these weaker acid components then displace even weaker acid components downstream to it, etc. Buffer components move down the column, with the successively stronger affinity (more acidic) components being retained, displacing weaker affinity (less acidic) components. Each component is effectively eluted off the column by the component having the next highest affinity, which is subsequently displaced by the component of next highest affinity, and so on. This succession of ampholyte displacement is termed as an "ampholyte train" by Murel et al.[29] This displacement process includes proteins which move down the column in bands of ampholyte that have the same affinity for the column anion-exchange sites as the proteins. A protein elutes at a pH that is determined by the acidity of the particular ampholyte band that it is in. Thus for anion chromatofocusing, the pH of the column is continuously lowered as successively more acidic components are eluted from the column, producing a continuously decreasing outlet pH gradient. Thus this mechanism also depicts proteins separated in the order of their pI values.

According to the displacement mechanism, the generation of a smooth linear pH gradient requires that there be an even distribution of mobile-phase components with incremental differences of affinities for the column's ion-exchange groups. This requirement is met by an even buffer capacity in the mobile phase throughout the gradient pH range, since the affinity value for a component, for the most part, correlates positively with its acid strength. Thus, both models require an even buffer capacity in the mobile phase (which in the displacement model results from an even distribution of ampholytes of successively differing affinities). However, unlike the buffer interaction model, the displacement model does not require the column to have an even buffering capacity throughout the gradient pH range. Thus strong ion-exchangers, which do not have buffering capacity in the required pH range for protein separation, have been occasionally used.[28,29] Also it was shown that Mono P columns, which are often used in chromatofocusing, have weak buffering capacity below a pH of 9, the range in which most separations take place in chromatofocusing.[8] These observations bring into question the presently held dogma in chromatofocusing which states that a column must have buffering capacity in the range of the generated pH gradient.

C. Mobile-phase Components and Considerations

Most conventional chromatofocusing techniques utilize elution buffer components from Amersham Biosciences. There are three elution buffers available: Polybuffer 74, Polybuffer 96, and Pharmalyte 8–10.5, designed for pH ranges 7–4, 9–6, and >9, respectively.[17,30] Generally, the largest pH range recommended is 3 pH units, although wide range techniques from pH 9.5 to 4.0[31] and from pH 8.5 to 3.5[32] have been reported. Narrow-range pH gradients (much less than a pH range of 3), however, yield the best resolution of proteins.[33–35] The appropriate application and elution buffers for a various pH ranges are given in several references.[17,18]

The exact chemical structures of the polyampholytes in commercial chromatofocusing elution buffers are proprietary information. These polyampholytes are poorly defined and vary in physical and chemical properties.[15] However, general chemical features of these components have been reported or can be reasonably surmised. It is widely assumed that the components are similar to carrier ampholytes used in IEF.[15] In fact, Polybuffers 74 and 96 have been shown to substitute effectively for standard carrier ampholytes in IEF techniques.[36] Polyampholytes used in IEF have also been used in chromatofocusing.[9] In the initial publication describing Polybuffers 74 and 96, the chemical structure of the various species in these polybuffers is reported to be amphoteric (meaning that they have both acidic and basic groups) with the buffering functionalities consisting principally of sterically protected tertiary amine groups, with a small percentage of sterically protected secondary amine groups and no primary amine groups.[30] It is reasonable to assume that the polybuffer species also contain carboxylic acid functionality(ies) to complete their amphoteric structure, in accordance with Vesterberg's report of "polyprotic amino carboxylic acids, each containing at least four weak protolytic groups, at least one being a carboxyl group and at least one a basic nitrogen atom, but no peptide bonds".[37,38] Vesterberg's work served as the foundation for the synthesis of Ampholine[38] (Amersham Biosciences) and presumably for other commercial ampholytes used in IEF. A structure for a hypothetical ampholyte component in Pharmalyte has been published showing both carboxylic acid groups and basic amines.[38] An extensive discussion on properties of various ampholytes used in IEF has been published.[38]

The exact requirements for mobile-phase buffers generating pH linear gradients on ion-exchange columns are not entirely clear. A requirement for a sufficiently high and even constant mobile-phase buffer capacity (or an even distribution of displacers of differing affinity) throughout the pH range of the gradient is widely acknowledged, being stipulated by both the buffer interaction and displacement mechanisms. This requirement is met by elution buffers containing multiple components having pI (or pK_a) values that span the pH range of the pH gradient.

What is unclear is whether or not there are restrictions on the chemical structures of the elution buffer components. Do considerations for the development of components used in IEF apply to chromatofocusing? The technique of IEF requires an amphoteric substance to serve as a carrier ampholyte, such that the ampholyte component (along with a protein having that exact same pI) will concentrate in the gel at a pH equal to its pI. Focusing of a particular ampholyte at its pH = pI location of the gel directly results from the amphoteric nature of the carrier ampholytes. For example, an ampholyte at the anode side of the neutral species (pH = pI) band will be pH < pI, resulting in its protonation and acquisition of a net positive charge. This positively charged ampholyte will be subsequently repelled by the positive anode back into the neutral ampholyte zone, where it again exists in a form having a net charge of zero. An analogous process take place for ampholytes on the cathodic side of the pI band, except in this case there is a loss of protons which gives a net negative charge to the ampholyte molecule. This results in its subsequent repulsion back into the neutral ampholyte zone by the cathode.

The development of a pH gradient on a chromatofocusing column is a substantially different process than IEF. First, it is not an equilibrium process in which a steady-state pH gradient is developed in the gel. On the contrary, chromatofocusing is a dynamic process in which pH is continually changing in the column. For another, there is one direction of movement in the chromatofocusing column (ignoring diffusion) toward the column outlet. This is different from IEF, in which there are two directions of movement depending on the charge properties of the molecules. Finally, there are significantly different roles for the stationary support in the two techniques. The proteins and mobile-phase components directly interact with the ion-exchange packing material in chromatofocusing, contributing significantly to the separation mechanism, while there is theoretically no interaction of these components with the gel in IEF. Thus, considerations for buffer components in IEF do not necessarily apply to the requirements of chromatofocusing.

There is not an a priori need, as in IEF, for amphoteric buffer species in chromatofocusing, since the development of pH gradients via buffer interaction or displacement mechanisms does not require the presence of more than one buffering functionality on a particular buffer component. However, in practice, there are only a few examples in which non-amphoteric elution buffers have been used in conventional chromatofocusing to generate internal pH gradients. Chromatofocusing studies have been published using a step change to an elution buffer containing multiple simple buffer components. The pH gradients generated from these attempts have not been smooth, instead show cascade steps, spikes, and protein plateau regions.[3,19,32,39,40] A possible cause for the pH spikes may be the retention of anionic buffer species on the anion-exchange column. These anionic species arise from the conversion of acidic components of the elution buffer to their conjugate base forms on exposure to the high pH column. These conjugate base anions are subsequently eluted as the pH decreases, causing the pH to spike. Frey and co-workers[26,41] addressed these difficulties by employing amine buffer components. These amine buffers generate smooth gradients because they are not retained by the anion-exchange column. However, polyampholyte solutions continue to be the elution buffer of choice in conventional chromatofocusing, due to their extensive track record of being able to generate reasonable pH gradients.

D. Column Components and Considerations for Anion Chromatofocusing

Examples of anion-exchange columns that have been used in anion-chromatofocusing experiments are given in Table 1, although other anion-exchange columns have been used.[9] The technique predominantly employs weak anion exchangers, although strong anion exchangers have also been used occassionally.[28,29] Important features of anion-exchange columns used for chromatofocusing are stability at high pH values, large pores (≥250 Å in pore diameter) to allow for the diffusion of proteins into the packing material, and use of either high-performance ion-exchange packing materials (10 μm particle diameter or less) for optimal results in terms of increased resolution, sensitivity, and speed of analysis (requires HPLC or FPLC pumps) or use of low-performance ion-exchange packing materials (>10 μm particle diameter) for less expensive instrumental components. The requirement of even buffering capacity of the column throughout the pH range of the gradient, as mentioned previously, may not be a strict requirement. However, current views still maintains a need for even buffering capacity for the column throughout the pH gradient range. Many published procedures use the columns specifically marketed for chromatofocusing by Amersham Biosciences, although there is no apparent advantage of these columns over other anion-exchange columns. There have been no real significant studies done comparing anion-exchange columns used in anion chromatofocusing. Specialized formats of chromatofocusing include the use of monolithic and micropellicular supports for separation of manganese peroxidase isoenzymes[42] and hemoglobin variants,[26] respectively, in less than 2 min. A microcolumn chromatofocusing technique (0.2 mm columns) is useful for analysis of small injection amounts.[43]

III. GRADIENT CHROMATOFOCUSING (EXTERNAL pH GRADIENT GENERATION)

Gradient chromatofocusing, developed by Anderson and co-workers,[6–8,44] uses a gradient HPLC system to generate an external pH gradient imposed on an anion-exchange HPLC column. This chromatofocusing technique is a more versatile approach to chromatofocusing than is the step-change method, having significant optimization and consumable expense advantages.

A. Description of Anion Gradient Chromatofocusing

1. Overview

The anion gradient chromatofocusing process is diagrammed in Figure 4. The column is first equilibrated with a high pH application buffer. Sample is then injected. A HPLC gradient system is then used to

TABLE I Examples of Anion Chromatofocusing Columns Used[15]

Column	Manufacturer	Low/high performance	Strong/weak anion-exchanger (group)	Support matrix	Particle diameter (μm)	Pore diameter (Å)	pH stability	Comments
Mono P	Amersham Biosciences	High	Weak (3° and 4° amines)	Polystyrene/ divinylbenzene	10		2–12	
PBE 94	Amersham Biosciences	Low	(3° and 4° amines)	Cross-linked agarose	40–120		3–12	Used in the pH range 4–9
PBE118	Amersham Biosciences	Low	(3° and 4° amines)	Cross-linked agarose	40–120		3–12	Used in the pH range 8–11
Bakerbond WP-PEI	J. T. Baker	High (and Low)	Weak (PEI[a])	Silica	5 [15, 40]	250–300	2–10	
SynChroPak AX-300 [AX-500] {AX-1000}	SynChrom	High	Weak (PEI[a])	Silica	6.5 and 7 [7] {10}	300 [500] {1000}	< 8.5	

[a]Polyethyleneimine.

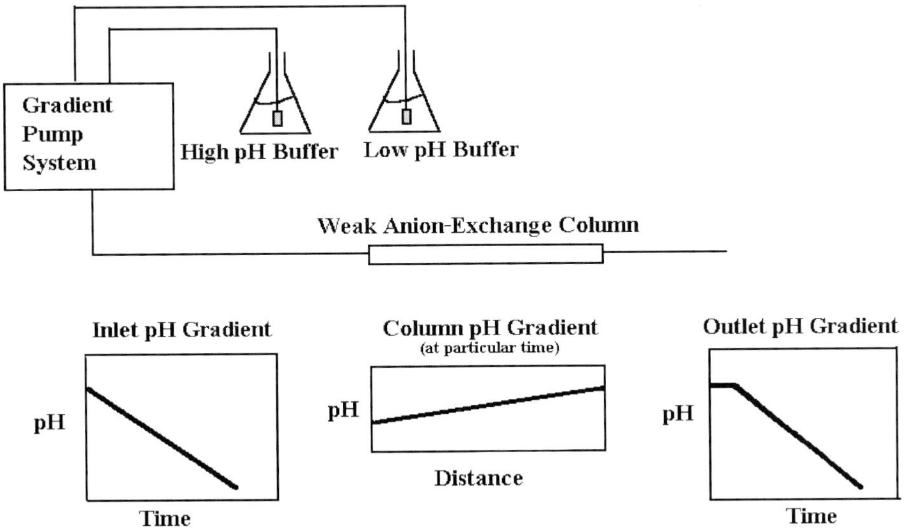

FIGURE 4 **Gradient anion chromatofocusing setup showing pH profile at various locations in the setup.**

produce a linear inlet pH gradient in time (produced by successively increasing the ratio of a low pH elution buffer to the high pH application buffer), which is introduced onto a high-performance anion-exchange column. This is different than the conventional technique which introduces a step change in pH to the column (see Figure 1). The inlet pH gradient is held up by the column, either through the column's buffering action or through a displacement mechanism, as discussed previously. This leads to the establishment of a pH gradient in distance in the column (internal pH gradient) and a pH gradient in time at the column outlet (outlet pH gradient). The focusing capability of gradient chromatofocusing is shown in Figure 5, which compares gradient chromatofocusing with salt gradient ion-exchange chromatography in analyzing the same sample with the same column. Figure 6 shows a comparable performance of the conventional and gradient chromatofocusing technique. However, gradient chromatofocusing can be further optimized as discussed below.

2. Mobile Phase and Column Components

An example application–elution buffer system used in gradient chromatofocusing is given in Table 2. The two buffers are mixed externally by a HPLC gradient system to generate a linear pH gradient from 7.6 to 4.[44] Note that the pK_a values for the various buffer components are spaced by 1 pH unit to give an even mobile-phase buffer capacity throughout the pH gradient. In theory, any weak anion-exchange column can be used

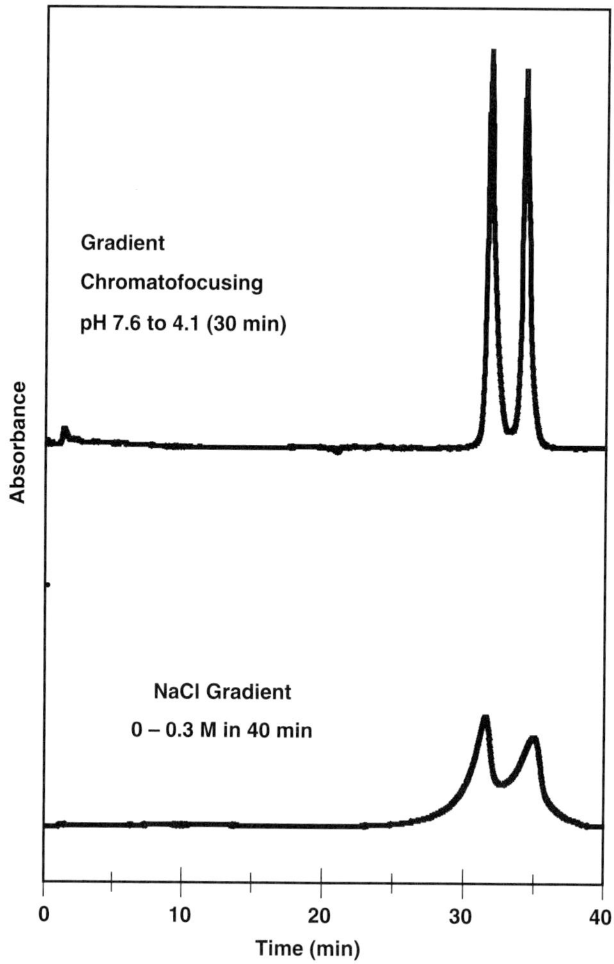

FIGURE 5 Gradient anion chromatofocusing compared with salt gradient anion-exchange chromatography in the analysis of β-lactoglobulin B (early eluting peak) and β-lactoglobulin A (late eluting peak). The same Mono P column and the same amount of injected protein sample were used for both studies. See Reference 8.

(and probably strong anion-exchange columns, although this has not yet been tested). Mono P (Amersham Biosciences) and Protein Pak DEAE (Waters) columns have been used.[6–8,44]

B. Advantages of Gradient Chromatofocusing over Conventional Chromatofocusing (Comparing the Anionic Techniques)

Gradient chromatofocusing has several distinct advantages over the conventional chromatofocusing technique. One limitation of the

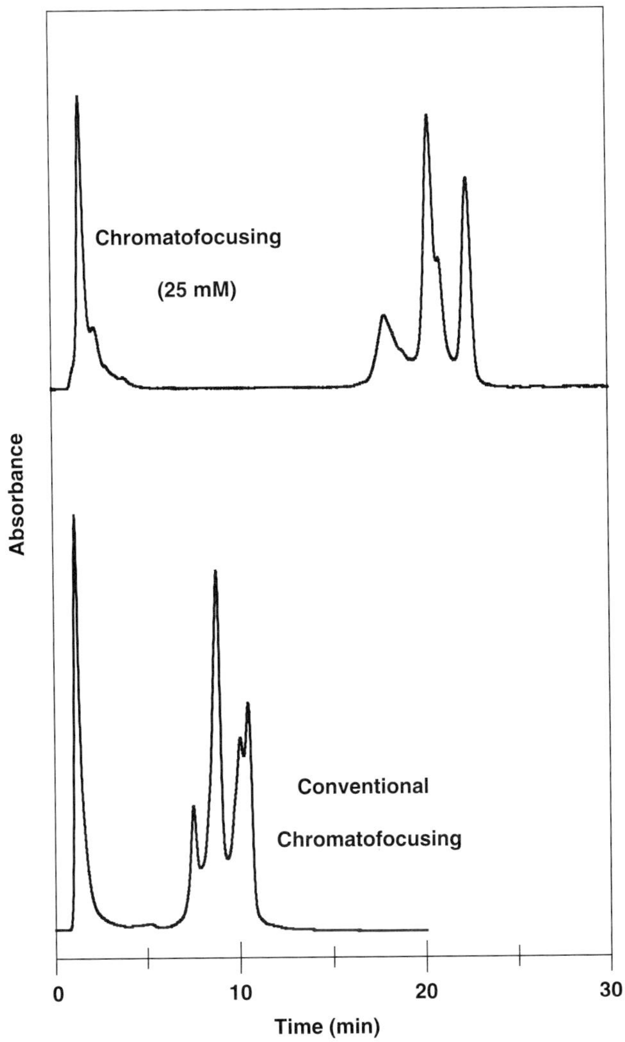

FIGURE 6 **Gradient chromatofocusing compared with conventional chromatofocusing in the analysis of the the same injected amount of a standard protein mixture (conalbumin, BSA, β-lactoglobulin B, ovalbumin, β-lactoglobulin A, in order of earliest to latest eluting peaks). The same Mono P column and pH gradient slope were used in both techniques. The earlier elution of the conventional technique is due to the gradient starting at a lower pH (same slope though). See Reference 8.**

conventional technique is that it utilizes polyampholyte buffers. Polyampholyte buffers are expensive, cause poor chromatographic reproducibility because of variability in their physical and chemical properties,[15] and associate with some proteins leading to additional complications.[45] As stated previously, attempts to use simple buffer components in conventional

TABLE 2 Example of Mobile-phase Buffers Used in Gradient Chromatofocusing[44]

	Application buffer (pK_a)	Elution buffer (pK_a)
Components	Bis-tris-propane (6.80) Piperazine (5.68)	Acetic acid (4.76) Lactic acid (3.81) Chloroacetic acid (2.87)
Concentrations of buffer used (for each component)	10–50 mM	6 to 50 mM
pH adjustment	Adjusted to pH 7.6 using HCl	No pH adjustment

technique have not been successful. A possible source of pH irregularities in the conventional technique, using simple buffer components, is the presence of acidic components in the elution buffer, which will bind in their conjugate base forms to anion-exchange sites as the elution buffer encounters the anion-exchange column equilibrated at high pH. In the conventional chromatofocusing design there is a step change to the elution buffer, which means that the concentration of acidic species exposed to high pH is at a maximum and thus a large amount of acidic species bind to the anion-exchange column. These species will elute later as the pH decreases, causing irregularities in the pH gradient.

Figure 4 illustrates the inherent advantage that the gradient chromatofocusing design has over conventional chromatofocusing by its ability to generate smooth linear pH gradients with simple buffer components. Gradient chromatofocusing minimizes irregularity effects by gradually introducing elution buffer to the column, meaning that any perturbation effects are also gradual. This leads to smoother gradients compared with pH gradients generated by a large change in pH. A further advantage of the design of gradient chromatofocusing in generating smooth gradients is its ability to adjust the gradient program to accommodate for any inherent irregularities in pH when buffers are mixed. For example, Shan and Anderson[8,44] utilized different proportioning rates throughout one gradient run to generate smooth reproducible linear pH gradients. This also gives gradient chromatofocusing the flexibility of generating any desired gradient shape.

Another limitation of the conventional chromatofocusing technique concerns the concentration of the buffer in the mobile phase. Relatively low buffer concentrations are required in the conventional technique for generating pH gradient slopes that are not too steep, in order to obtain reasonable resolution.[1,2,19,46] Hearn and Lyttle[19] found that a buffer concentration range of 2.5–5 mM Buffalyte WR 3-10 (Pierce chemical Company) was needed to produce suitable pH gradients for conventional chromatofocusing analysis. A high concentration of buffer in the mobile phase produces too steep a pH gradient slope in the conventional technique, as the mobile-phase buffering capacity overwhelms the buffering capacity of the column. Gradient chromatofocusing can use a wide range of buffer concentrations, from very low to as high as 75 mM concentration of each buffer component.[8] This capability of employing high buffer concentrations results from its design feature in which incremental pH changes are introduced to the column via the external gradient, not a precipitous drop in pH. There is no influence of buffer concentration on the outlet pH gradient slope in gradient chromatofocusing (although concentration does influence the slope of the internal pH gradient, with lower mobile-phase buffer concentrations giving steeper internal pH gradients),[8,44] allowing the use of 10–20 times higher concentration buffers than is used in the conventional technique. Use of higher concentration buffers is highly advantageous for solubilization of proteins, which can be a problem in conventional chromatofocusing. It has also been shown that buffer concentration is an important independent parameter in separation, with resolution gains at higher buffer concentrations of 3- to 25-fold noted compared with the conventional technique.[8] The effect of buffer concentration on the separation of proteins is shown in Figure 7.

Finally, there is a distinct advantage of gradient chromatofocusing in its ease of varying the outlet pH gradient slope through manipulation of the gradient program; steep slopes are advantageous for speed of analysis and shallow gradients are advantageous for high resolution. Figure 8 shows the increased resolution for proteins obtained when shallow pH gradients are used. Thus, gradient chromatofocusing addresses a deficiency of the conventional technique in generating shallow pH gradients.[16]

IV. PERFORMANCE CHARACTERISTICS

A. pI Determination

A comparison from various studies of the elution pH in conventional anion chromatofocusing versus pI determined by IEF is given in Table 3.[47–57] The difference between the two techniques (elution pH minus IEF pI) is plotted in Figure 9. The differences range from +1.5 to −1.7 pH units, having an absolute average value of 0.76 pH units. Thus chromatofocusing provides, at best, a rough approximation of pI. For this reason the pH of elution in chromatofocusing is referred to as an "apparent pI." Table 3 shows an almost equal number of positive deviations (13 of 30) as negative deviations (17 of 30) and an average difference close to zero, indicating that there is no directional bias for chromatofocusing compared with IEF. Figure 9 indicates that there is also no appreciable pI-dependent trend in the difference (elution pH minus IEF pI), as the plot has a very small correlation coefficient ($r^2 = 0.027$).

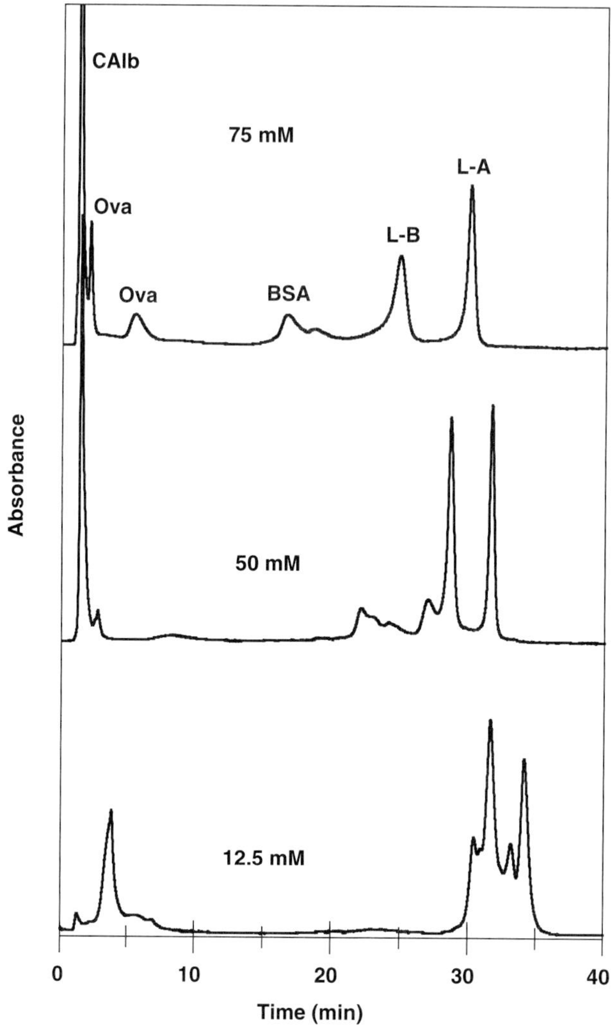

FIGURE 7 **Gradient chromatofocusing studies showing the effect of using different concentrations of buffer components in the application and elution buffers in the analysis of a standard protein mixture (conalbumin, BSA, β-lactoglobulin B, ovalbumin, β-lactoglobulin A, in order of earliest to latest eluting peaks). Buffer concentrations are the concentrations for each component in the application (bis-tris propane and piperazine) and elution (pyridine, acetic acid, lactic acid, and chloroacetic acid) buffers. See Reference 8.**

There are several reasons for the difference between a protein's chromatofocusing elution pH and its IEF pI. One significant reason is that the retention of a protein on an ion-exchange column is through the interaction of only a portion of the protein's structure with the ion exchanger; proteins usually bind through their strongest interacting sites.[58,59] Thus, it is not the

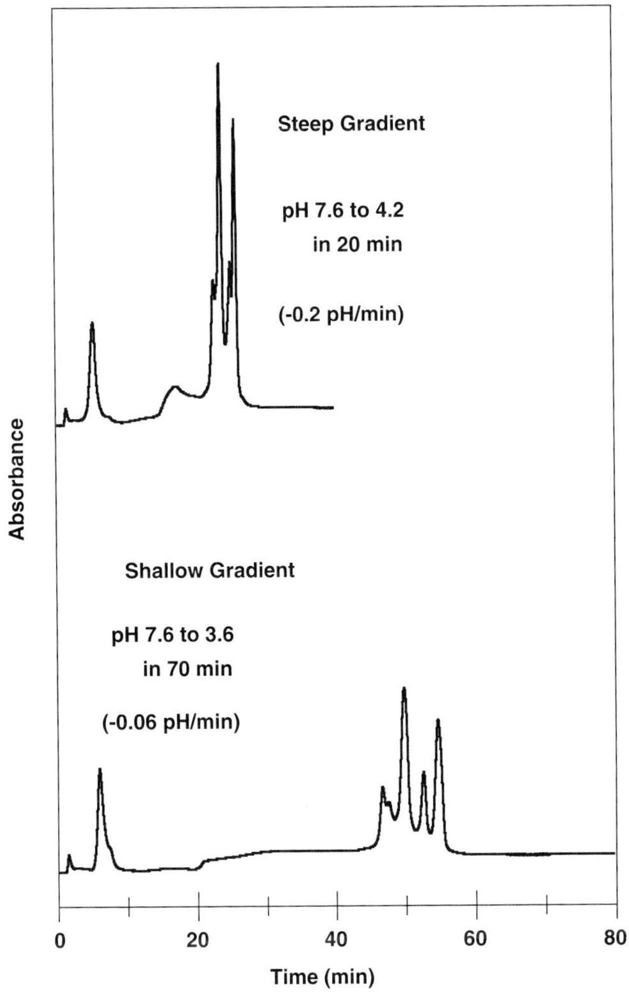

FIGURE 8 **Gradient chromatofocusing studies showing increased resolution for a standard protein mixture (conalbumin, BSA, β-lactoglobulin B, ovalbumin, and β-lactoglobulin A, in order of earliest to latest eluting peaks) when the pH gradient slope is decreased.**

charge state of the entire protein, but rather the charge state of the interacting surface region of the protein that influences its retention and elution. It would not be expected that a technique dependent on a portion of the protein's structure (chromatofocusing) would be an accurate indicator of a property arising from the entire protein structure (pI). The elution pH of a protein in chromatofocusing could be higher or lower than its pI, since there is no predictable relationship between the acid/base properties of the protein's surface regions and that of the entire molecule.

TABLE 3 Summary of Studies Comparing Elution pH of Conventional Anion Chromatofocusing Using Ampholyte Elution Buffers and IEF pI

Protein	Elution pH chromatofocusing	pI determined by IEF	Difference (elution pH-IEF value)	Reference
Albumin fraction 5 (bovine)	4.69	4.90	−0.21	47
Biliverdin reductase-1 (rat)	6.91	6.23	0.68	48
Biliverdin reductase-2 (rat)	6.33	5.91	0.42	48
Creatine kinase MM-1 (human heart and skeletal muscle)	8.3	7.2	1.1	49
Creatine kinase MM-2 (human heart and skeletal muscle)	7.9	6.9	1	49
Creatine kinase MM-3 (human heart and skeletal muscle)	7.6	6.7	0.9	49
Creatine kinaseMM-4 (human heart and skeletal muscle)	7.2	6.4	0.8	49
Creatine kinaseMM-5 (human heart and skeletal muscle)	6.8	6.2	0.6	49
Glutathione S-Transferase-A1 (human hepatic)	5.4	7.1	−1.7	50
Glutathione S-Transferase-C1 (human hepatic)	9.0	7.5	1.5	50
Glutathione S-Transferase-C2 (human hepatic)	8.7	7.3	1.4	50
25-hydroxyvitamin D3 binding protein (rat)	4.58	4.93	−0.35	51
Insulin (bovine)	4.88	5.72	−0.84	47
β-lactamase (class C) (Pseudomonas aeruginosa)-1	8.0	8.7	−0.7	52
β-lactamase (class C) (Pseudomonas aeruginosa)-2	7.7	8.3	−0.6	52
β-lactamase (class C) (Pseudomonas aeruginosa)-3	7.6	8.2	−0.6	52
β-lactamase (class C) (Pseudomonas aeruginosa)-4	7.3	7.6	−0.3	52
β-lactoglobulin A (ovine)	5.2	5.8	−0.6	53
β-lactoglobulin A (ovine)	4.8	5.8	−1.0	53
β-lactoglobulin A	4.34	5.14	−0.8	47
β-lactoglobulin B	4.48	5.31	−0.83	47
Leucine enkephalin	5.61	5.50	0.11	47
Methionine-enkephalin	5.58	5.40	0.18	47

TABLE 3 (*Continued*)

Protein	Elution pH chromatofocusing	pI determined by IEF	Difference (elution pH-IEF value)	Reference
Myoglobin-1 (sheep heart)	7.9	7.4	0.5	54
Myoglobin-2 (sheep heart)	8.1	7.16	0.94	54
Pore-forming protein (murine lymphocyte)	4.7	6.4	−1.7	55
S-hexylglutathione-binding protein (30 kDa) (human hepatocellular carcinoma)	4.9	5.8	−0.9	56
Soybean agglutinin isolectin-1	6.7	7.0	−0.3	57
Soybean agglutinin isolectin-2	6.3	6.85	−0.55	57
Soybean agglutinin isolectin-3	6.0	6.7	−0.7	57
		Average of Absolute Values	0.76	
		Average (keeping sign)	−0.085	
		SD	0.87	

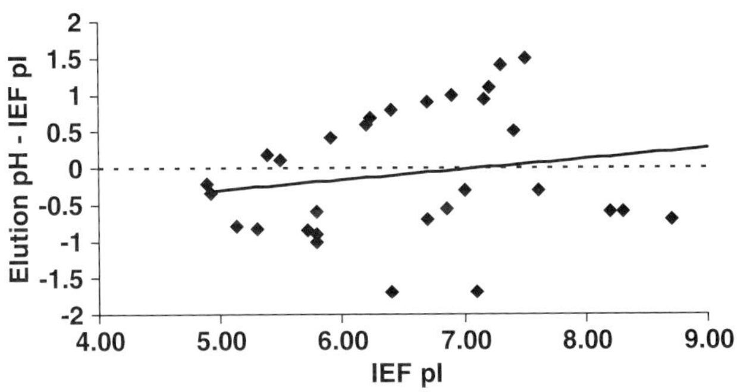

FIGURE 9 **Plot of (elution pH minus IEF pI (y value)) versus (IEF pI (*x* value)) for various proteins determined in studies comparing anion-chromatofocusing elution pH and IEF pI (from Table 3). The least-squares regression line (solid line) through the data is plotted and is $y = 0.1417x - 1.011$ with $r^2 = 0.026812$.**

Another possible cause for the discrepancy of the elution pH compared with the IEF pI is the Donnan potential of the ion exchanger which leads to different equilibrium pH values in the same column segment for the mobile and stationary phases. The buffer interaction model assumes that these pH values are the same.[1,2] The Donnan potential, whose magnitude decreases and then reaches a plateau with increasing mobile-phase buffer concentration,[2] causes the pH of the mobile phase to be less than the pH of the stationary phase in a particular column segment (this is for anion-exchange columns; the opposite is observed for cation-exchange columns). Therefore, when the stationary phase reaches the pH = pI of the protein, the protein changes to a net zero charge and is desorbed into the mobile phase. The mobile phase is at a lower pH than the corresponding stationary-phase segment. Thus, the protein will elute from the column at this lower mobile phase pH. However, the higher pH of the stationary-phase segment from which the protein originally desorbed is a truer reflection of the protein's pI. Thus, for anion chromatofocusing, there is a decrease in elution pH from the protein's pI value due to the Donnan potential.

Other possible reasons for deviations from pI are a mixed retention mechanism on ion-exchangers, such as hydrophobic retention[60,61] and anion concentration elution effects.[8]

B. Peak Widths

Peak widths at half-height for proteins have been reported as 0.06–0.09 pH units for the conventional technique and 0.02–0.15 pH units for the gradient technique.[8] Peak width is dependent on the pH gradient slope and buffer concentration. Shallower pH gradients produce the narrowest peaks widths (pH units) in gradient chromatofocusing.[8] Greater focusing of protein peaks (in terms of pH width) occurs at low buffer concentrations and high Donnan potentials in conventional chromatofocusing.[1,2] Decreased buffer concentration leads to narrowing of the protein peak (pH units) by decreasing the slope of the pH gradient and by increasing the Donnan potential of the ion exchanger.[1] The Donnan potential can also be increased by using high-capacity ion exchangers.[1] It should be realized, however, that the more shallow the pH gradient the more the dilution of the protein peak, since it is eluted over a longer time period. Theoretically, peak width is independent of flow rate,[2] as has been experimentally verified.[2] However, another study showed decreased resolution with increased flow rate.[46]

V. APPLICATIONS

Chromatofocusing is frequently used in protein purification as one in a series of chromatographic steps. Chromatofocusing is often used in conjunction with one or more other non-denaturing chromatographic techniques in protein purification. The other chromatographic techniques include gel filtration (size-exclusion), regular salt-elution ion exchange,

TABLE 4 Separation of Multiple Protein Forms and Multiple Proteins by Conventional Chromatofocusing

β-D-N-acetylglucosaminidase (lysosomal isozymes A, I1, I2, B)[62]	IFN-α (human, 5 peaks)[90]
Acid phosphatase (rat) (prostate, 8 isoforms; seminal 5 isoforms)[63]	Interleukin-2 (human, 4 peaks)[91]
α-fetoprotein (bovine, 7 isoforms; human, 3 isoforms)[64]	Interleukin-1 β[92]
Apolipoproteins (human HDL, 8 peaks)[65]; (ApoB, ApoA-I, ApoC-II)[66]; (human VLDL, 7 peaks)[67,68]; (human VLDL, C-II, C-III0, C-III1, C-III2)[69]	β-Lactamase (*Bacillus cereus*, I, II, and others)[93]
Creatine kinase (MM isoforms)[70]	Luteinizing hormone (ovine, 13 isoforms[94]; human, 3 peaks[95])
Crystalline subunits, lens[71]	Phospholipase A (rat liver, lysosomal, 5 isoenzymes)[96]
Cytochrome P-450 (mouse liver, 8 isoforms[72]; rat liver, 21 fractions[73])	Prolactin (rat, 5 variants)[97]
Ferritin (rabbit liver isoforms)[74]	Prostate-specific antigen (free PSA (several isoenzymes) and complexed PSA)[98]
Follicle-stimulating hormone (human isoforms[75]; rat, 11 components[76])	Serum proteins[99]
Galactosyltransferase (human, 11 peaks)[77]	Steroid receptor proteins (estrogen receptors)[100,101]
Glutathione S-transferase (rat brain, 12 forms[78]; human liver, 13 forms[79]; rat liver, 7 isoenzymes[80])	Thyroid-hormone-binding globulin (human, isoforms)[102]
Glyceraldehyde-3-phosphate dehydrogenase (Jaculus orientalis, muscle, 3 isoforms)[81]	Thyroid-stimulating hormone (human, 4–6 peaks;[103] ovine, ≥9 isohormones[104]; bovine deglycosylated, ≥3 peaks[104])
Hemoglobin (human, metHb, 2 Hb valence intermediates, oxygenated Hb[82]; human S, A0, F0, F1[83])	Transferrin (human, carbohydrate-deficient glycoprotein syndrome, 3 isoforms)[105]
Hexokinase (yeast isoenzymes PI and PII)[84]	UDP-glucuronosyltransferase (human liver, 2 isoenzymes[106]; rat liver, isoforms[107])
Human chorionic gonadotropin (isoforms)[85]	Virus capsid proteins (foot-and-mouth disease, VP1, VP2, VP3, VP0)[108]
Immunoglobulins (murine monoclonal Fab'[86]; human IgA, IgM, IgG[87]; clonotypic antibodies[88]; IgG[89])	

hydroxylapatite, affinity, and hydrophobic interaction chromatography. Chomatofocusing is also used to separate multiple protein samples and multiple forms of proteins. Published examples of the separation of multiple proteins or multiple forms of a protein by chromatofocusing are given in Table 4.[62–108] Many of the isoforms listed in Table 4 are different glycoforms of the proteins. Chromatofocusing studies characterizing glycoforms of proteins include follicle-stimulating hormone glycoforms,[109] human chorionic gonadotropin glycoforms,[110] liver and hepatoma sialoglycoproteins,[111] and various sialoglycoproteins.[112] Other general types of separations accomplished by chromatofocusing include the separation/ purification of phosphorylated proteins,[113–116] iodinated peptides,[117] deamidated forms of proteins,[118] and membrane proteins.[119,120]

Chromatofocusing has been used in several proteomic studies. Lubman and co-workers[121–124] have been on the forefront of this application of chromatofocusing. In these studies, two-dimensional chromatographic analysis was done, in which chromatofocusing is used in a first dimension pI-based fractionation (pI characterization), with each fraction then subsequently separated on a second-dimension reversed-phase column (hydrophobicity characterization). This was then followed by a mass spectrometric analysis (molecular weight characterization) of the separated proteins. The technique was used in the profiling of human breast cancer whole cell lysates,[121] comparing the protein expression map of virulent and non-virulent *E. coli*,[122] and comparing the protein expression of untreated and drug-treated human colon adenocarcinoma cells.[123] It has also been employed by Lubman and co-workers to make protein microarrays, in which fractions from the two-dimensional chromatographic separation are deposited on nitrocellulose slides subsequently used to characterize humoral response in cancer.[124] In a different proteomic study, approximately 125 proteins from *Haemophilus influenzae* were identified in chromatofocusing fractions.[125] The two-dimensional technique has been commercialized by Beckman-Coulter as their ProteomeLab™ PF 2D system.

REFERENCES

1. Sluyterman, L. A. A. and Elgersma, O. Chromatofocusing: isoelectric focusing on ion-exchange columns. I. General principles. *J. Chromatogr.* 150:17–30, 1978.
2. Sluyterman, L. A. A. and Wijdenes, J. Chromatofocusing: isoelectric focusing on ion-exchange columns. II. Experimental verification. *J. Chromatogr.* 150:31–44, 1978.
3. Sluyterman, L. A. A. and Wijdenes, J. Chromatofocusing. III. The properties of a DEAE-agarose anion exchanger and its suitability for protein separations. *J. Chromatogr.* 206:429–440, 1981.
4. Sluyterman, L. A. A. and Wijdenes, J. Chromatofocusing. IV. Properties of an agarose poly(ethyleneimine) ion exchanger and its suitability for protein separations. *J. Chromatogr.* 206:441–447, 1981.
5. Leaback, D. H. and Robinson, H. K. Ampholyte displacement chromatography – a new technique for the separation of proteins illustrated by the resolution *β*-*N*-acetyl-D-hexosaminidase isoenzymes unresolvable by isoelectric focusing or conventional ion-exchange chromatography. *Biochem. Biophys. Res. Commun.* 67:248–254, 1975.
6. Liu, Y. and Anderson, D. J. Gradient chromatofocusing high-performance liquid chromatography. I. Practical aspects. *J. Chromatogr. A* 762:207–217, 1997.
7. Liu, Y. and Anderson, D. J. Gradient chromatofocusing high-performance liquid chromatography. II. Theoretical aspects. *J. Chromatogr. A* 762:47–54, 1997.

8. Shan, L. and Anderson, D. J. Gradient chromatofocusing. Versatile pH gradient separation of proteins in ion-exchange HPLC: characterization studies. *Anal. Chem.* 74:5641–5649, 2002.
9. Ivanov, A. V. and Nesterenko, P. N. Generation and application of pH gradients in liquid chromatography. *J. Anal. Chem.* (transl. *Zh. Anal. Khim.*) 54:494–510, 1999.
10. Giri, L. Chromatofocusing. In *Guide to Protein Purification* (*Methods Enzymol.*, 182), (Deutscher, M. P., Ed.) Academic Press, San Diego, pp. 380–392, (1990).
11. Sluyterman, L. A. A. and Kooistra, C. Ten years of chromatofocusing: a discussion. *J. Chromatogr.* 470:317–326, 1989.
12. Mohammad, J. Chromatofocusing. In *Purifying Proteins for Proteomics* (Simpson, R. J., Ed.) Cold Spring Harbor Laboratory Press, Woodbury, NY, pp. 355–379, 2004.
13. Mantle, T. J. and Noone, P. Chromatofocusing. In *Protein Purification*, 2nd ed., (*Methods in Mol. Biol.*, 244), pp. 233–237, 2004.
14. Lukačin, R. and Deppert, W. R. Chromatofocusing. In *Protein Liquid Chromatography* (*J. Chromatogr. Libr.* 61), (Kastner, M., Ed.) Elsevier, Amsterdam, pp. 385–414, 2000.
15. Hutchens, T. W. Chromatofocusing. In *Protein Purification: Principles, High-Resolution Methods, and Applications*, 2nd ed. (Janson, J.-C. and Rydén, L., Eds.) W-VCH, New York, pp. 207–238, 1998.
16. Li, C. M. and Hutchens, T. W. Chromatofocusing. In *Practical Protein Chromatography* (*Methods in Mol. Biol*, 11) (Kenney, A. and Fowell, S., Eds.) Humana Press, Totowa, NJ, pp. 237–248, 1992.
17. *Chromatofocusing with Polybuffer and PBE Handbook*, Amersham Pharmacia Biotech, 18-1140-62.
18. At Amersham Biosciences website www4.amershambiosciences.com. Articles are entitled: "Buffer systems for broad pH intervals in chromatofocusing" and "Buffer systems for narrow pH intervals in chromatofocusing".
19. Hearn, M. T. W. and Lyttle, D. J. Buffer-focusing chromatography using multicomponent electrolyte elution systems. *J. Chromatogr.* 218:483–495, 1981.
20. Frey, D. D., Barnes, A. and Strong, J. Numerical studies of multicomponent chromatography using pH gradients. *AIChE J.* 41:1171–1183, 1995.
21. Frey, D. D. Local-equilibrium behavior of retained pH and ionic strength gradients in preparative chromatography. *Biotechnol. Prog.* 12:65–72, 1996.
22. Strong, J. C. and Frey, D. D. Experimental and numerical studies of the chromatofocusing of dilute proteins using retained pH gradients formed on a strong-base anion-exchange column. *J. Chromatogr. A* 769:129–143, 1997.
23. Bates, R. C. and Frey, D. D. Quasi-linear pH gradients for chromatofocusing using simple buffer mixtures: local equilibrium theory and experimental verification. *J. Chromatogr. A* 814:43–54, 1998.
24. Narahari, C. R., Strong, J. C. and Frey, D. D. Displacement chromatography of proteins using a self-sharpening pH front formed by adsorbed buffering species as the displacer. *J. Chromatogr. A* 825:115–126, 1998.
25. Bates, R. C., Kang, X. and Frey D. D. High-performance chromatofocusing using linear and concave pH gradients formed with simple buffer mixtures. I. Effect of buffer composition on the gradient shape. *J. Chromatogr. A* 890:25–36, 2000.
26. Kang, X. and Frey, D. D. Chromatofocusing using micropellicular column packings with computer-aided design of the elution buffer composition. *Anal. Chem.* 74:1038–1045, 2002.
27. Emond, J. P. and Pagé, M. An approach to ampholyte displacement chromatography. *J. Chromatogr.* 200:57–63, 1980.
28. Murel, A., Vilde, S., Pank, M., Shevchuk, I. and Kirret, O. Chromatophoresis: a new approach to the theory and practice of chromatofocusing. I. General principles. *J. Chromatogr.* 347:325–334, 1985.
29. Murel, A., Vilde, S., Pank, M., Shevchuk, I. and Kirret, O. Chromatophoresis: a new approach to the theory and practice of chromatofocusing. II. Experimental verification. *J. Chromatogr.* 362:101–112, 1986.

69. Huff, M. W. and Strong, W. L. P. Separation and isolation of human apolipoproteins C-II, C-III0, C-III1, and C-III2 by chromatofocusing on the fast protein liquid chromatography system. *J. Lipid Res.* 28:1118–1123, 1987.
70. Abendschein, D. R., Fontanet, H. L. and Nohara, R. Optimized preservation of isoforms of creatine kinase MM isoenzyme in plasma specimens and their rapid quantification of semi-automated chromatofocusing. *Clin. Chem.* 36:723–727, 1990.
71. Body, P. and Bloemendal, H. Simultaneous separation of all lens crystallin subunits by chromatofocusing. *FEBS Lett.* 232:39–45, 1988.
72. Marriage, H. J. and Harvey, D. J. Resolution of mouse hepatic cytochrome P-450 isozymes by chromatofocusing. *J. Chromatogr.* 354:383–392, 1986.
73. Pasanen, M. and Pelkonen, O. Resolution of multiple P-450 forms: separation of aryl hydrocarbon hydroxylase and aminopyrine N-demethylase-associated P-450 isoenzymes by chromatofocusing. *Gen. Pharmacol.* 16:361–366, 1985.
74. Frenkel, E. J., Van den Beld, B., Konig, B. W. and Marx, J. J. M. Preparative isolation of distinct molecular forms of rabbit liver ferritin using high-performance liquid chromatography. *Anal. Biochem.* 135:489–494, 1983.
75. Walton, W. J., Nguyen, V. T., Butnev, V. Y., Singh, V., Moore, W. T. and Bousfield, G. R. Characterization of human FSH isoforms reveals a nonglycosylated β-subunit in addition to the conventional glycosylated β-subunit. *J. Clin. Endocrinol. Metab.* 86: 3675–3685, 2001.
76. Blum, W. F. P., Riegelbauer, G. and Gupta, D. Heterogeneity of rat FSH by chromatofocusing: studies on in-vitro bioactivity of pituitary FSH forms and effect of neuraminidase treatment. *J. Endocrinol.* 105:17–27, 1985.
77. Barbe, D., Gauduchon, P., Bar, E., Heron, J. F. and Le Talaer, J. Y. Isoenzymic patterns of human galactosyltransferase from normal and cancer patients. *Cancer Detect. Prev.* 8:279–285, 1985.
78. Thyagaraju, K., Devi, K.N., Rao, A.D., Balaji, M., Hildenbrandt, G.R., Ramamurthi, R. and Reddy, C.C. Purification and characterization of 12 forms of rat brain glutathione s-transferases. *J. Parasitol. Appl. Anim. Biol.* 5:33–48, 1996.
79. Jagt, D. L. V., Hunsaker, L. A., Garcia, K. B. and Royer, R. E. Isolation and characterization of the multiple glutathione S-transferases from human liver. Evidence for unique heme-binding sites. *J. Biol. Chem.* 260:11603–11610, 1985.
80. Jensson, H., Aalin, P. and Mannervik, B. Glutathione transferase isoenzymes from rat liver cytosol. *Methods Enzymol.* 113 (*Glutamate, Glutamine, Glutathione, Relat. Compd.*), 504–507, 1985.
81. Soukri, A., Valverde, F., Hafid, N., Elkebbaj, M. S. and Serrano, A. Characterization of muscle glyceraldehyde-3-phosphate dehydrogenase isoforms from euthermic and induced hibernating *Jaculus orientalis. Biochim. Biophys. Acta* 1243:161–168, 1995.
82. Bolzacchini, E., Fermo, I., Rovida, E., Colombo, R. and Samaja, M. Separation of the valence intermediates of human hemoglobin by high-performance chromatofocusing. *J. Chromatogr.* 397:233–237, 1987.
83. Alexander, N. M. and Neeley, W. E. Chromatofocusing of human hemoglobins. *J. Chromatogr.* 230:137–141, 1982.
84. Jacob, L., Beecken, V., Bartunik, L. J., Rose, M. and Bartunik, H. D. Purification and crystallization of yeast hexokinase isoenzymes: characterization of different forms by chromatofocusing. *J. Chromatogr.* 587:85–92, 1991.
85. Diaz-Cueto, L., Barrios-de-Tomasi, J., Timossi, C., Mendez, J. P. and Ulloa-Aguirre, A. More in-vitro bioactive, shorter-lived human chorionic gonadotropin charge isoforms increase at the end of the first and during the third trimesters of gestation. *Mol. Hum. Reprod.* 2:643–650, 1996.
86. Tarburton, J. P. and Halpern, S. E. Chromatofocusing studies involving a monoclonal Fab'. *J. Nucl. Med.* 33:2148–2153, 1992.
87. Waldrep, J. C. and Schulte, J. R. HPLC chromatofocusing of human immunoglobulins. *J. Immunol. Methods* 118:273–277, 1989.

88. Endo, Y., Miyai, K., Hata, N. and Iijima, Y. Effect of pH on antigen binding by clonotypic antibodies with different isoelectric points. *Biotechnol. Appl. Biochem.* 9:74–81, 1987.
89. Gallo, P., Olsson, O. and Siden, A. Small-column chromatofocusing of cerebrospinal fluid and serum immunoglobulin G. *J. Chromatogr.* 375:277–283, 1986.
90. Bodo, G. and Adolf, G. R. Formation of human IFN-alpha subtype mixtures by different human cells. *Contrib. Oncol.* 20 (*Physiol. Pathol. Interferon Syst.*) 134–141, 1984.
91. Gerard, J. P. and Bertoglio, J. Chromatofocusing as a tool for the characterization and partial purification of human interleukin-2. *J. Immunol. Methods* 55:243–251, 1982.
92. Wingfield, P. T., Graber, P., Rose, K., Simona, M. G. and Hughes, G. J. Chromatofocusing of N-terminally processed forms of proteins. Isolation and characterization of two forms of interleukin-1 beta and of bovine growth hormone. *J. Chromatogr.* 387:291–300, 1987.
93. Gal, S., Toth-Martinez, B. L. and Kiss, L. Use of chromatofocusing for separation of β-lactamases. III. β-Lactamases of the *Bacillus cereus* 569/H strain. *J. Chromatogr.* 264:170–173, 1983.
94. Zalesky, D. D., Nett, T. M. and Grotjan, H. E. Ovine luteinizing hormone: isoforms in the pituitary during the follicular and luteal phases of the estrous cycle and during anestrus. *J. Anim. Sci.* 70:3851–3856, 1992.
95. Snyder, P. J., Bashey, H. M., Gatewood, C. V. and Karowe, M. Characterization of human LH isohormones from fresh pituitary tissue. *Mol. Cell. Endocrinol.* 54:115–121, 1987.
96. Hostetler, K. Y., Yazaki, P. J. and Van den Bosch, H. Purification of lysosomal phospholipase A. Evidence for multiple isoenzymes in rat liver. *J. Biol. Chem.* 257:13367–13373, 1982.
97. Briski, K. P., Swanson, G. N. and Sylvester, P. W. Size and charge heterogeneity of pituitary and plasma prolactin in the male rat. *Neuroendocrinology* 63:437–445, 1996.
98. Huber, P. R., Schmid, H.-P., Mattarelli, G., Strittmatter, B., Van Steenbrugge, G. J. and Maurer, A. Serum free prostate specific antigen: Isoenzymes in benign hyperplasia and cancer of the prostate. *Prostate* 27:212–219, 1995.
99. Young, J. L. The separation of serum proteins by chromatofocusing. *Protides Biol. Fluids* 30:611–614, 1982.
100. Hutchens, T. W., Dunaway, H. E. and Besch, P. K. High-performance chromatofocusing of steroid receptor proteins in the presence and absence of steroid. Investigation of steroid-dependent alterations in surface charge heterogeneity. *J. Chromatogr.* 327:247–259, 1985.
101. Boyle, D. M., Wiehle, R. D., Shahabi, N. A. and Wittliff, J. L. Rapid, high-resolution procedure for assessment of estrogen receptor heterogeneity in clinical samples. *J. Chromatogr.* 327:369–376, 1985.
102. Terasaki, T. and Pardridge, W. M. Differential binding of thyroxine and triiodothyronine to acidic isoforms of thyroid hormone binding globulin in human serum. *Biochemistry* 27:3624–3628, 1988.
103. Pickles, A. J., Peers, N., Robertson, W. R. and Lambert, A. Different isoforms of human pituitary thyroid-stimulating hormone have different relative biological activities. *J. Mol. Endocrinol.* 9:251–256, 1992.
104. Keel, B. A., Harms, R. L. and Amir, S. M. The charge characterization of native and deglycosylated thyrotropin. *Endocr. Res.* 16:151–163, 1990.
105. Yamashita, K., Ideo, H., Ohkura, T., Fukushima, K., Yuasa, I., Ohno, K. and Takeshita, K. Sugar chains of serum transferrin from patients with carbohydrate deficient glycoprotein syndrome. Evidence of asparagine-N-linked oligosaccharide transfer deficiency. *J. Biol. Chem.* 268:5783–5789, 1993.
106. Irshaid, Y. M. and Tephly, T. R. Isolation and purification of two human liver UDP-glucuronosyltransferases. *Mol. Pharmacol.* 31:27–34, 1987.
107. Chowdhury, J. R., Chowdhury, N. R., Falany, C. N., Tephly, T. R. and Arias, I. M. Isolation and characterization of multiple forms of rat liver UDP-glucuronate glucuronosyltransferase. *Biochem. J.* 233:827–837, 1986.

108. Murdin, A. D., Doel, T. R. and Spier, R. E. Isolation of capsid proteins of foot-and-mouth disease virus by chromatofocusing. *J. Virol. Methods* 7:207–216, 1983.
109. Creus, S., Chaia, Z., Pellizzari, E. H., Cigorraga, S. B., Ulloa-Aguirre, A. and Campo, S. Human FSH isoforms: carbohydrate complexity as determinant of in-vitro bioactivity. *Mol. Cell. Endocrinol.* 174:41–49, 2001.
110. Yoshimura, M., Pekary, A. E., Pang, X. P., Berg, L., Goodwin, T. M. and Hershman, J. M. Thyrotropic activity of basic isoelectric forms of human chorionic gonadotropin extracted from hydatidiform mole tissues. *J. Clin. Endocrinol. Metab.* 78:862–866, 1994.
111. Karaivanova, V., Ivanov, S. and Chelibonova-Lorer, K. Pattern of sialoglycoproteins obtained by chromatofocusing of chicken liver and hepatoma MC-29 microsomal preparations labeled in vivo with 3H-leucine and N-acetyl-14C-mannosamine. *Cancer Biochem. Biophys.* 12:275–282, 1992.
112. Burness, A. T. H. and Pardoe, I. U. Chromatofocusing of sialoglycoproteins. *J. Chromatogr.* 259:423–432, 1983.
113. Boesze-Battaglia, K., Kong, F., Lamba, O. P., Stefano, F. P. and Williams, D. S. Purification and light-dependent phosphorylation of a candidate fusion protein, the photoreceptor cell peripherin/rds. *Biochemistry* 36:6835–6846, 1997.
114. Adamus, G., Arendt, A., Hargrave, P. A., Heyduk, T. and Palczewski, K. The kinetics of multiphosphorylation of rhodopsin. *Arch. Biochem. Biophys.* 304:443–447, 1993.
115. Aton, B. R. Illumination of bovine photoreceptor membranes causes phosphorylation of both bleached and unbleached rhodopsin molecules. *Biochemistry* 25:677–680, 1986.
116. Aton, B. R., Litman, B. J. and Jackson, M. L. Isolation and identification of the phosphorylated species of rhodopsin. *Biochemistry* 23:1737–1741, 1984.
117. Woloszczuk, W. Iodogen-catalyzed iodination of human calcitonin and Tyr(0)-katacalcin and purification of their mono- and di-iodinated derivatives by chromatofocusing. *J. Immunol. Methods* 90:1–6, 1986.
118. Oray, B., Yuksel, K. U. and Gracy, R. W. Separation of deamidated forms of triosephosphate isomerase by chromatofocusing. A comparison of chromatofocusing with column isoelectric focusing. *J. Chromatogr.* 265:126–130, 1983.
119. Lin, J. T., Schwarc, K. and Stroh, A. Chromatofocusing and centrifugal reconstitution as tools for the separation and characterization of the sodium cotransport systems of the brush-border membrane. *Biochim. Biophys. Acta* 774:254–260, 1984.
120. Wakefield, L. M., Cass A. E. and Radda, G. K. Isolation of a membrane protein by chromatofocusing: cytochrome b-561 of the adrenal chromaffin granule. *J. Biochem. Biophys. Methods* 9:331–341, 1984.
121. Chong, B. E., Yan, F., Lubman, D. M. and Miller, F. R. Chromatofocusing nonporous reversed-phase high-performance liquid chromatography/electrospray ionization time-of-flight mass spectrometry of proteins from human breast cancer whole cell lysates: a novel two-dimensional liquid chromatography/mass spectrometry method. *Rapid Commun. Mass Spectrom.* 15:291–296, 2001.
122. Zheng, S., Schneider, K. A., Barder, T. J. and Lubman, D. M. Two-dimensional liquid chromatography protein expression mapping for differential proteomic analysis of normal and O157:H7 *Escherichia coli. BioTechniques* 35:1202–1208, 1210–1212, 2003.
123. Yan, F., Subramanian, B., Nakeff, A., Barder, T. J., Parus, S. J. and Lubman, D. M. A comparison of drug-treated and untreated HCT-116 human colon adenocarcinoma cells using a 2-D liquid separation mapping method based upon chromatofocusing PI fractionation. *Anal. Chem.* 75:2299–2308, 2003.
124. Yan, F., Sreekumar, A., Laxman, B., Chinnaiyan, A. M., Lubman, D. M. and Barder, T. J. Protein microarrays using liquid phase fractionation of cell lysates. *Proteomics* 3:1228–1235, 2003.
125. Fountoulakis, M., Langen, H., Gray, C. and Takacs, B. Enrichment and purification of proteins of *Haemophilus influenzae* by chromatofocusing. *J. Chromatogr. A* 806:279–291, 1998.

13 ALTERNATIVE ELECTROFOCUSING METHODS

CORNELIUS F. IVORY

Department of Chemical Engineering, Washington State University, Pullman, WA 99164-2710

I. INTRODUCTION

A scientist setting up a core proteomics facility today would have little choice but to use isoelectric focusing (IEF) as one of the orthogonal steps in a multidimensional separations cascade preceding final analysis by mass spectrometry (MS). However, as more complex organisms with

Handbook of Isoelectric Focusing and Proteomics
D. Garfin and S. Ahuja, editors.

a greater diversity of proteins and a larger dynamic range are analyzed, the shortcomings of two-dimensional polyacrylamide gel electrophoresis (2-D-PAGE) will coax scientists to move their protocols out of large-format, cross-linked gels and into multiscale, networked channels filled with viscous fluids, e.g., buffered polymer solutions.

Over the next few years the field will witness a shift to the chromatographies, which offer a cornucopia of orthogonal chemistries, straightforward sizing and relatively easy automation, but suffer from an Achilles' heel, i.e., they require both length and time to resolve peaks. To be specific, it may be difficult to adapt liquid chromatography to the short, microfluidic columns needed for the last fractionation dimensions of low-abundance proteins (LAP) while maintaining the peak capacities needed for systeomics. In addition to IEF which is influenced less by length, alternative electrofocusing methods (AFMs), which do not focus solutes at their isoelectric points (pI), may allow scientists to circumvent this limitation.

A. Equilibrium-gradient Methods

The AFMs analyzed in this chapter are part of a subset of the equilibrium gradient methods (EGM; Figure 1) described by Giddings and Dahlgren,[1] which use an applied electric field or electric field gradient as at least one of the counterbalanced forces on a focused solute. A complete binary set of EGMs would consist of dozens of pairs of forces, some of which are mentioned by Giddings[2] in the context of field-flow fractionation (FFF) as well as myriad variations on isocratic and gradient-elution chromatographies, each paired against a second force, e.g., hydrodynamic. It is unlikely that all possible binary pairs have been discovered, much less exploited, at this time and it does not appear that any ternary EGMs have been reported in the literature to date, so this area may still be considered "immature" and therefore ripe for further development. The motivation for this chapter is to describe what is known about this emerging set of methods in the hope that "gadgeteers" and theorists will explore this frontier and thereby help scientists develop new tools for systems biology.

As an aside, the reader should note that the use of the word "equilibrium" in EGM is not warranted since these systems are actually either in a steady state or a pseudosteady state, i.e., when the peak is moving but not changing shape. Still, Giddings' intent is made clear in his mathematical treatment of EGMs.

B. Alternative Electrofocusing Methods

Perhaps the most important instance of field-gradient electrofocusing was O'Farrell's invention[3] of counter-acting chromatographic electrophoresis (CACE) (Figure 2), in which he demonstrated that a protein

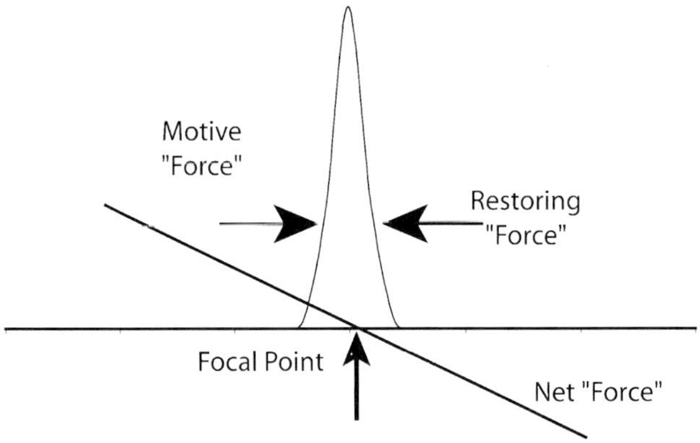

FIGURE 1 AFMs are a subset of the EGMs in which solutes are driven to a unique focal point by applying a counteracting restoring gradient against a constant motive "force." Adapted from Giddings and Dahlgren.[1]

could be focused to concentrations in excess of 100 mg/mL at the interface between two size-exclusion chromatography (SEC) gels by directing the electrophoretic migration of that protein against the step change in the chromatographic flow which occurred at that interface. To do this he placed several centimeters of low molecular weight cutoff (MWCO) SEC packing, Bio-Rad P-10 polyacrylamide gel, above a length of high MWCO SEC packing. Bio-Rad A-50 m agarose gel, directed the flow of 10 mM Tris acetate buffer at pH ~7.4 down through these packings at a nominal superficial velocity, $U \sim 0.44$ cm/min, and directed a nominal electric field, $E \sim 12$ V/cm, so that the target solute, ferritin, and other proteins would migrate against the flow of buffer.[4]

Since the protein could not enter the P-10, it moved through the packing with the interstitial velocity, U/ε, where the void fraction ε is typically about 0.4 in soft gels. In the lower part of the column the protein could access virtually all of the interior volume of the A-50 m and would move at a lower apparent velocity $\sim U$. This opened a range of values of the applied electric field $U/\mu < E < U/\varepsilon\mu$, where μ is the electrophoretic mobility, which would allow the protein to collect at the interface between the two gels.

Shortly thereafter, several groups published theoretical analyses[5] and experimental extensions[6,7] of this work, gradually elucidating the mechanism which allowed electrophoretic focusing without the use of pH gradients. Perhaps the most startling result to arise from the experimental work was that the focused protein concentrations routinely exceeded 100 mg/mL and, in some experiments, appear to have approached 300 mg/mL in the interstitial space between the P–10 beads. The implication here was that, since CACE did not require the protein to focus at

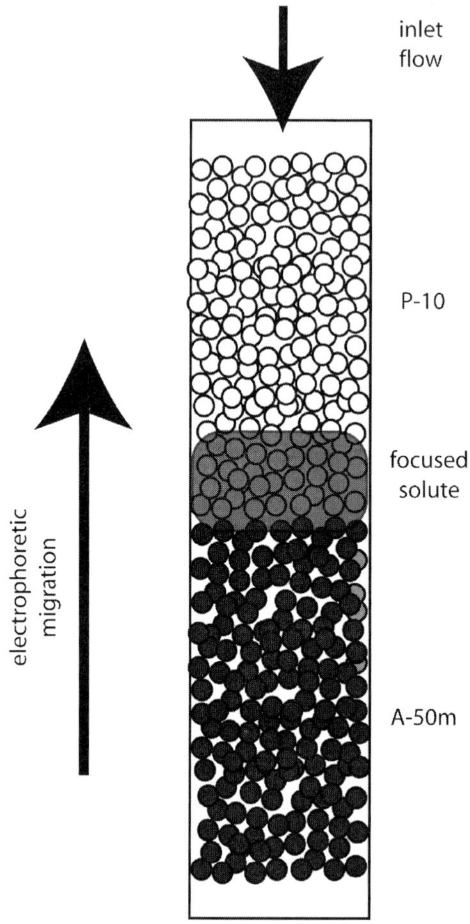

FIGURE 2 CACE focuses a protein above the interface between two SEC packings where the upstream packing has a low K_{AV} (accessible volume coefficient) while the downstream packing has a large K_{AV}. The accessible volume coefficient is the fraction of a chromatographic packing that a solute can occupy. This opens up a window, i.e., a range of operating parameters, within which the protein has a focal point at the interface. Adapted from Gobie and Ivory.[7]

its pI where solubility is generally lowest or in a very low conductivity ampholyte buffer, it could avoid the isoelectric precipitation that plagues IEF of complex biological solutions. The absence of a pH gradient also allowed that non-amphoteric solutes, e.g., some metabolites, and polynucleic acids, which depurinate at their pI, could be focused.

The most important result from the theoretical analyses of that time was that CACE shared a common mathematical basis with IEF and that these techniques were part of an as yet undiscovered family of electrofocusing techniques[8] which would, by mathematical association, also be members of the set of EGMs delineated by Giddings.

II. THEORY

Presenting a complete theory of electrophoresis that would embrace all potential electrofocusing EGMs is beyond the scope of this chapter. A treatment that covers IEF in detail is available in Mosher's *Dynamics of Electrophoresis*[9] and in a series of subsequent papers[10–13] by these authors as well as others[14–17] on this topic. However, to understand the underlying principles involved in the AFMs, one only needs to consider the component of the mass flux equation which is parallel to the electric field,

$$N_{x,i} = -D_i \frac{dC_i}{dx} + (U_{x,i} + \mu_i E_x)C_i = 0 \tag{1}$$

where the electric field points along the x-axis. $N_{x,i}$ is the mass flux in the x direction, i.e., the mass flow through a channel per unit area, C_i is the solute concentration of the focusing species, $U_{x,i}$ is the chromatographic velocity of the solute, μ_i is its electrophoretic mobility, and E_x is the x-component of the electric field. D_i is a dispersion coefficient that has the same units as a diffusion coefficient but which accounts for the effects of the convective velocity profile on band-broadening, e.g., Taylor dispersion.[18,19] Equation (1) accounts, in mathematical terms, for the different mechanisms by which the solute can move about in the channel: diffusion, convection, and electrophoresis.

If the solute is focused and its concentration profile has reached a steady state, then since it is not moving in space, the solute flux is zero. The first term on the right-hand side of Equation (1) is a dispersive term which does not affect the location of the focused band but which does affect its breadth. A dispersive term is used in place of molecular diffusion because the focusing solutes move at different speeds depending on how close they are to fixed surfaces and this variation in speed often increases peak spreading. The terms inside the parentheses determine whether the solute will focus and they set the location of the focused band at the point where the sum of those terms vanishes. In order for a solute to focus at the unique point where the sum of terms in parentheses vanishes, at least one of the terms in parentheses must vary, usually in a monotonic gradient of some sort, and the slope of that gradient must force each solute toward its unique focal point.[20] In physical terms this simply means that, if a solute is moved away from its focal point, the forces acting on it will drive it back to that point. If the slope of the gradient is wrong, then these forces would tend to *defocus* or spread the solute, a technique which might be useful for diluting or mixing solutes.

If we replace the electric field with the current density divided by the electrical conductivity, i.e., $E_x = i_x/\sigma$, and the electrophoretic mobility by charge times the absolute mobility, i.e., $\mu_i = z_i\omega_i$, then

$$N_{x,i} = -D_i\frac{\mathrm{d}C_i}{\mathrm{d}x} + \left(U_{x,i} + z_i\omega_i\frac{i_x}{\sigma}\right)C_i = 0 \tag{2}$$

the flux equation explicitly shows five different parameters in parentheses that can be adjusted in order to focus charged solutes. The most familiar of these, IEF is a case where the fluid velocity is identically zero, $U_{x,i} = 0$, and the charge z_i which is a function of pH, vanishes at the pI values of the proteins. The term in the parentheses vanishes at the pI and this location becomes the focal point for a given protein. CACE corresponds to the case where the electrical term is constant but the chromatographic velocity $U_{x,i}$ varies. As will be seen later, other variations on this theme hold the velocity constant against a gradient in the electric field.

Even though the same fundamental equation (2) can be used to describe each of these different electrofocusing techniques, they yield subtly different expressions for the variance and the resolution that may give an advantage to one or the other in different situations. In the next section, the mathematical relations for the variance and resolution of several different real and hypothetical electrofocusing devices are derived and then compared to illustrate the relative advantages and disadvantages of each technique.

The reader should keep in mind that the linear model, Equation (2), does not take into account non-linear, electrostatic coupling among the various ionic species in solution. Linear theory specifically ignores the constraint of local electroneutrality and, instead, assumes that the electric field is determined by the electrode voltages, geometry, and the bulk conductivity of the electrolyte in the separation channel while ignoring the response of the electrolyte ions to the electric field. Because of this, it is likely that other phenomena, which are overlooked in the linear theory, may allow the development of different protocols based on these or other parameters. For example, Equation (2) would not allow prediction of isotachophoresis (ITP) although it is likely that each of the AFMs can exhibit ITP-like behavior which would allow solutes to "stack" at high concentrations.

A. Generalized Theory

While Equation (2) can be used to describe most electrofocusing techniques, it can also be generalized to describe other EGMs like density-gradient centrifugation or gradient-elution chromatography. To do this, Equation (2) is rewritten in the generalized form

$$N_{x,i} = -D_i\frac{\mathrm{d}C_i}{\mathrm{d}x} + (-S_i x + I_i)C_i = 0 \tag{3}$$

where S_i is a positive number which represents the *slope* of the sum of the "forces" acting on the solute and I_i is an *intercept* parameter which sets the location of the gradient. These two parameters may be varied independent of each other. Then the slope will determine the width of the focused peak and the resolution of any pair of peaks while the intercept will determine where in our channel the peak(s) will focus.

Equation (3) may be integrated using separation of variables[21] and the resulting constant of integration may be determined by applying the integral constraint that a known mass M_T of solute is focused in a chamber with cross-sectional area A_c to get

$$C_i = \frac{M_T}{A_c}\sqrt{\frac{S_i}{2\pi D_i}}\exp\left(-\frac{S_i(x-x_{f,i})^2}{2D_i}\right). \tag{4}$$

Extracting spatial moments[22] from Equation (4) using the formula

$$m_{n,i} = \int_{-\infty}^{\infty} x^n C_i \, dx \tag{5}$$

yields the location $x_{f,i} = m_{1,i}/m_{0,i} = I_i/S_i$ and spatial variance $\chi_i^2 = m_{2,i}/m_{0,i} - X_{f,i}^2 = D_i/S_i$ of the peak. $m_{n,i}$ is the mathematical definition of the nth peak moment; $m_{0,i}$ is the zeroth moment of the peak concentration distribution and is a measure of the mass of solute in the focused peak; $m_{1,i}$ is the first peak moment and, when normalized by the zeroth moment, provides the mean or mass-average location of the peak; and $m_{2,i}$ is the second peak moment which is used in the above formula to calculate the *spatial* variance of the peak about its mean location. The square root of the variance, i.e., the standard deviation, is a measure of the width of the peak which is used in the formula below for the resolution.

The doubly infinite range of the moment integrals, $(-\infty, \infty)$, is used since this provides a simple result that is an excellent approximation for each moment as long as the edges of the peak do not extend past the ends of the focusing apparatus. This simplification yields an approximate form for the resolution (Figure 3) of two solutes:

$$R_s \approx \frac{|x_{f,1} - x_{f,2}|}{1/2(W_1 + W_2)} \approx \frac{|x_{f,1} - x_{f,2}|}{2(\chi_1 + \chi_2)} \approx \frac{|\Delta I\langle S\rangle - \Delta S\langle I\rangle|}{\sqrt{4D_i\langle S\rangle^3}} \approx \frac{|\Delta I|}{\sqrt{4D_i\langle S\rangle}} \tag{6}$$

where the shorthand for the averages is $\langle S\rangle = (S_1+S_2)/2$, $\langle I\rangle = (I_1+I_2)/2$, for the differences is $\Delta S = (S_1-S_2)$, $\Delta I = (I_1-I_2)$, and the two assumptions have been made that (1) the dispersion coefficients for both solutes are equal and (2) $\langle S\rangle >> \Delta S$. The expression on the far right corresponds to the ideal case where the gradient is constant everywhere so that $\Delta S = 0$. From this we see the classic result that the resolution is inversely proportional to the square root of the gradient for all EGMs described by flux equations that have the same form as Equation (3).

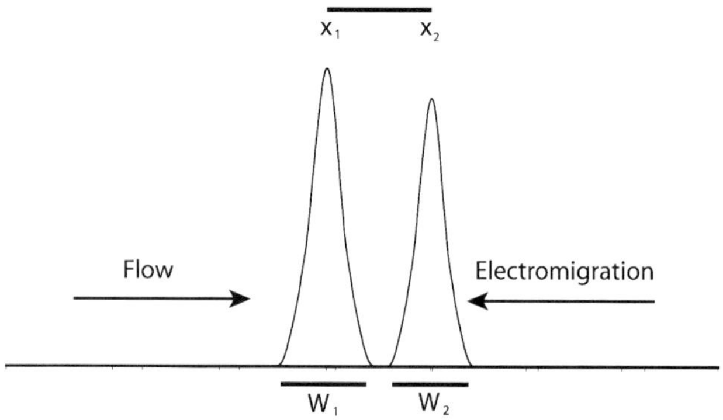

FIGURE 3 **Resolving power of a focusing method can be estimated from Equation (6), where $\Delta x = x_2 - x_1$ is the distance between two stationary peaks and W_1, W_2 are the baseline widths of those peaks.**

FIGURE 4 **Isoelectric focusing in a PDMS MEMS channel. About 1 ng each of three naturally fluorescent proteins were loaded into a 2-cm-long separation channel and focused in broad-range (3–10) ampholyte. The channel width from bottom to top is 300 μm and its depth is 5 μm. 1, allo-phycocyanin; 2, phycoerythrin, 3–5, green fluorescent protein.**

B. IEF

This same approach can be applied to IEF (Figure 4) under the assumptions that the pH profile is linear and is known while the conductivity is constant and the velocity U is zero. Although there is a good deal of published work on *non-linear* IEF, which produces a more rigorous result, linearization allows us to compare each of the different AFMs on a common basis. Starting with Equation (2), we expand the charge z_i in a Taylor series in the pH around its pI value and expand the pH in a series around the focal point $x_{f,i}$. This yields

$$z_i = (\mathrm{pH} - \mathrm{pI}_i)\left.\frac{\mathrm{d}z_i}{\mathrm{d(pH)}}\right|_{\mathrm{pI}_i} \tag{7a}$$

$$\mathrm{pH} = \mathrm{pI}_i + (x - x_{f,i})\left.\frac{\mathrm{d(pH)}}{\mathrm{d}x}\right|_{x_{f,i}} \tag{7b}$$

which, when combined into Equation (2), has a structure similar to Equation (3):

$$-D_i\frac{dC_i}{dx}+\left((x-x_f)\frac{d(pH)}{dx}\frac{dz_i}{d(pH)}\frac{\omega_i i_x}{\sigma}\right)C_i=0 \tag{8}$$

where the peak is located at its pI and the spatial variance is found to be

$$\chi^2=\frac{\sigma D_i}{\frac{d(pH)}{dx}\frac{dz_i}{d(pH)}\omega_i i_x} \tag{9}$$

From these formulas, IEF resolution can be expressed as

$$R_{IEF}=\frac{|\Delta pI|}{4}\sqrt{\frac{\frac{\omega_i i_x}{\sigma D_i}\frac{dz_i}{d(pH)}}{d(pH)/dx}} \tag{10}$$

which is equivalent to the classic result given by Righetti.[23]

Note that the resolution increases with increasing current density and decreases with increasing dispersion and/or conductivity. In fact, one of the reasons that PAGE gels often produce better results than capillaries is that the cross-linked gels reduce the diffusion coefficient of the solute in the gel more than they reduce its mobility.

Finally, if the pH gradient is held constant in space, as would be the case with Immobilines®, and a constant flow U is applied parallel to this gradient, e.g., as a result of electroosmosis, then this theory predicts that the target solutes will focus some distance away from their characteristic pI values.

C. Velocity-gradient Focusing

Two extant examples of velocity-gradient focusing, CACE and the SepStack,[24] both rely on step changes in the apparent velocity U_i so they are not readily comparable with the other processes treated above which use continuous gradients. However, one can postulate a long channel in which one or more surfaces is a membrane, e.g., ultrafiltration (UF), which transmits solvent and small ions but retains larger target solutes (Figure 5). Assuming that a constant flux J_w of buffer is drawn out of the separation channel though the membrane, and that a constant electric field E_x is directed against the axial flow of buffer, a range of values of E_x will exist which will cause the target solute to focus within the channel.

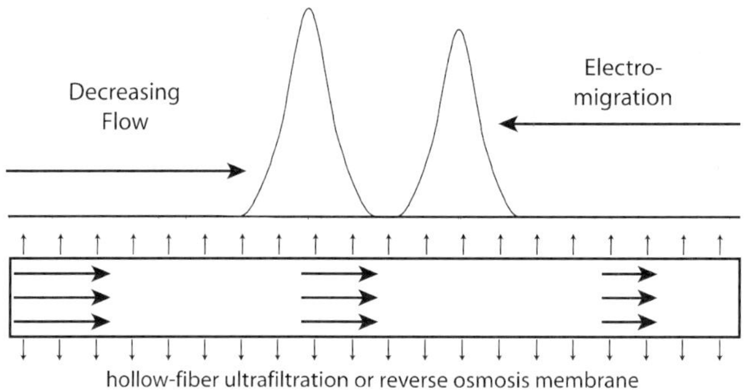

FIGURE 5 Velocity-gradient focusing, which is closely related to CACE, can be carried out in a hollow-fiber membrane where liquid is allowed to pass through the membrane, creating a gradient in the axial flow. Solutes are then focused using a fixed electric field as the counteracting force.

This electrofocusing system can be described by the flux equation

$$N_{x,i} = -D_i \frac{dC_i}{dx} + \left(\frac{Q_{in}}{wh} - \frac{J_w x}{h} + \mu_i E_x \right) C_i = 0 \qquad (11)$$

where Q_{in} is the volumetric flow introduced into a rectangular channel and w and h are the width and height of the separation channel respectively; the membrane covers the width and length of the lower surface of the channel.

In analogy with Equation (3), this model yields a spatial variance $\chi_i^2 = hD_i/J_w$ and resolution:

$$R_{VU} = \frac{|\Delta\mu|}{4} \sqrt{\frac{hE_x^2}{DJ_w}} \qquad (12)$$

It is worth noting from Equation (12) that the resolution has an unusually strong electric field dependence and that the permeate flux J_w can be made arbitrarily small to improve resolution.

D. Electric Field-gradient Focusing

There are a number of different ways in which the electric field or current density can be directly manipulated to yield a gradient, but the two that have been explored to date are (1) by shaping (Figure 6) the geometry of the electrode channel[25,26] or (2) by using an array of individually adjustable electrodes.[27] Assuming that the separation channel is operated with a linear electric field, the flux can be expressed as

$$N_{x,i} = -D_i \frac{dC_i}{dx} + \left(\frac{Q_{in}}{wh} - x\mu_i \frac{dE_x}{dx} + \mu_i E_x|_{x=0} \right) C_i = 0 \qquad (13)$$

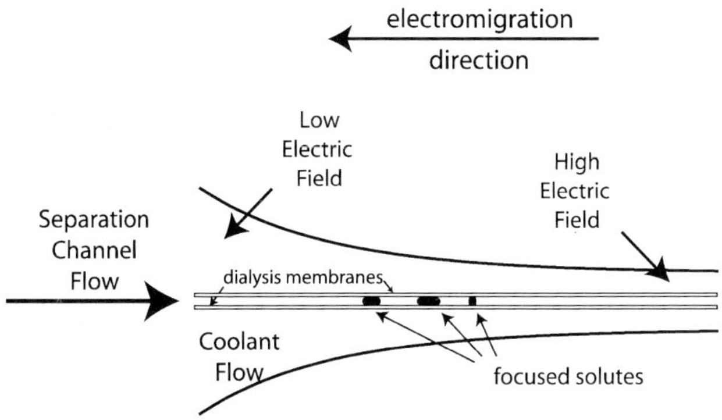

FIGURE 6 EFGF was originally performed in a chamber with a fluted purge or coolant channel. The separation channel has a fixed cross-sectional area and is in contact with the coolant channel by means of a membrane which passes current but does not pass fluid, e.g., a dialysis membrane. The Achilles' heel of the EFGF methods is the membrane which may be difficult to install, especially in microscale channels. Adapted from Koegler and Ivory.[25]

from which, by analogy with Equation (3), the focal point, spatial variance and the resolution are

$$x_{\mathrm{f}} = \frac{(Q_{\mathrm{in}}/w\mathrm{h}) + \mu_i E_x|_{x=0}}{\mu_i[\mathrm{d}E_x/\mathrm{d}x]} \tag{14a}$$

$$\chi^2 = \frac{D_i}{\mu_i}\left[\frac{\mathrm{d}E_x}{\mathrm{d}x}\right]^{-1} \tag{14b}$$

$$R_{\mathrm{EFGF}} = \frac{|\Delta\mu|}{4\langle\mu\rangle}\frac{(Q_{\mathrm{in}}/w\mathrm{h})}{\sqrt{\langle\mu\rangle_i D_i[dE_x/\mathrm{d}x]}} \tag{14c}$$

respectively. Here as before, the electric field gradient dE/dx may be made arbitrarily small to increase the resolution.

E. Conductivity-gradient Focusing

Although simpler to set up, and run than EFGF, conductivity-gradient focusing (CGF or electromobility focusing (EMF)) is considerably more difficult to analyze mathematically because the salt gradient profile must first be determined and then used to estimate the electric field gradient. Consider a separation chamber composed of two channels, a separation channel with height h, and a dialyzate channel with height d, each with width w, and separated by a membrane which passes current and electrolytes (Figure 7), but not liquid, e.g., a dialysis membrane. The dialyzate channel is rapidly flushed with low-conductivity buffer σ_d and so its conductivity is effectively constant, but the high-conductivity buffer metered into the separation channel is allowed to diffuse through the membrane, setting up a gradient that runs from the inlet conductivity to the purge conductivity in a non-linear fashion. The change in conductivity in the separation channel can be expressed as

$$\frac{Q_{\mathrm{in}}}{w\mathrm{h}}\frac{\mathrm{d}\langle\sigma(x)\rangle}{\mathrm{d}x} = -\frac{D_\sigma}{\mathrm{h}^2}(\langle\sigma(x)\rangle - \sigma_d) \quad \text{where } \langle\sigma(x)\rangle = \frac{1}{\mathrm{h}}\int_0^h \sigma(x, y)\,\mathrm{d}y \tag{15}$$

where $\langle\sigma\rangle$ is the conductivity averaged over the height of the separation channel. Equation(15) integrates to

$$\langle\sigma(x)\rangle = (\langle\sigma\rangle_0 - \sigma_d)\mathrm{e}^{-(wD_\sigma/hQ_{\mathrm{in}})x} + \sigma_d \tag{16}$$

where $\langle\sigma\rangle_0$ is the inlet conductivity at $x = 0$ in the separation channel.

The total current $I_{\mathrm{T},x}$ is free to distribute itself between the two compartments and, since current will follow the path of least resistance between the separation and dialyzate channels, it generates an electric

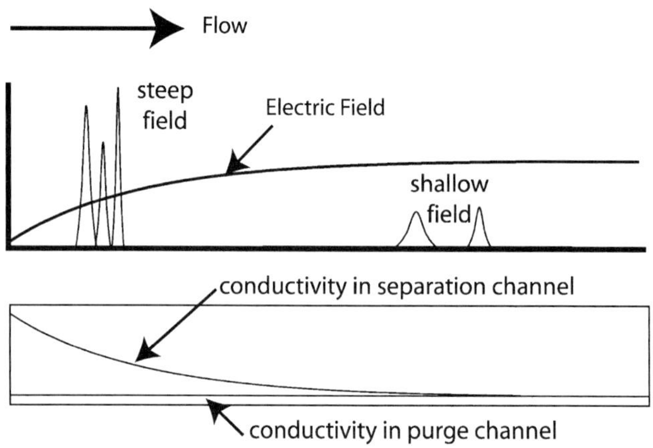

FIGURE 7 **CGF uses mismatched conductivities between the separation channel and the purge or coolant channel to create an electric-field gradient. In this example, the coolant conductivity is small and constant while the separation channel inlet conductivity is large and varies exponentially over the length of the channel. Solute peaks focused near the inlet are sharp but poorly resolved while peaks focused nearer the outlet are generally better resolved.**

field in the form

$$E_x = \frac{I_{\mathrm{T},x}}{\langle\sigma(x)\rangle w\mathrm{h} + \sigma_d wd} \tag{17}$$

Plugging this expression into the flux equation

$$N_{x,i} = -D_i \frac{\mathrm{d}C_i}{\mathrm{d}x} + \left(\frac{Q_{\mathrm{in}}}{w\mathrm{h}} - \frac{\mu_i I_{\mathrm{T},x}}{\langle\sigma(x)\rangle w\mathrm{h} + \sigma_d wd}\right) C_i = 0 \tag{18}$$

yields a differential equation for the concentration profile which has the focal point

$$x_{\mathrm{f}} = -\frac{\mathrm{h}Q_{\mathrm{in}}}{wD_\sigma} \ln\left[-\frac{\mathrm{h}\mu_i I_{\mathrm{T},x} + (d+\mathrm{h})\sigma_d Q_{\mathrm{in}}}{\mathrm{h}(\langle\sigma(0)\rangle + \sigma_d) Q_{\mathrm{in}}}\right] \tag{19}$$

where, in order for x_{f} to be a positive number so that our solute will focus within the channel, the range of operation of the focusing channel must be

$$(d+\mathrm{h})\sigma_d < \left|\frac{\mathrm{h}\mu_i I_{\mathrm{T},x}}{Q_{\mathrm{in}}}\right| < \left(\mathrm{h}\langle\sigma(0)\rangle + d\sigma_d\right) \tag{20}$$

Although it is possible to derive mathematical forms for the spatial variance and the resolution, the resulting expressions are unwieldy and simpler, approximate forms can be derived by expanding the term in parentheses in a Taylor series about the focal point to obtain

$$N_{x,i} = -D_i \frac{\mathrm{d}C_i}{\mathrm{d}x} - \alpha_i (x - x_{\mathrm{f}}) C_i = 0$$

$$\text{where } \alpha_i = \frac{(\mathrm{h}\mu_i I_{\mathrm{T},x} + (d+\mathrm{h})\sigma_d Q_{\mathrm{in}})}{\mathrm{h}^3 \mu_i I_{\mathrm{T},x}} D_\sigma \tag{21}$$

which has spatial variance $\chi_i^2 = D_i/\alpha_i$, and resolution

$$R_{\nabla\sigma} = \frac{|x_{\mathrm{f},1} - x_{\mathrm{f},2}|}{4} \frac{\alpha_1 \alpha_2}{\sqrt{\mathrm{D}\langle\alpha\rangle^3}} \tag{22}$$

While this formula is difficult to express in a way that makes it directly comparable with others, it can still be used as a basis for comparison when values of the separation parameters are supplied (see Table 1).

F. Temperature-gradient Focusing

A recent addition to the AFMs is temperature-gradient focusing[28,29] in which an axial temperature gradient is used to provide a gradient in

TABLE 1 Theoretical Resolution and Peak Capacity of Alternative Focusing Techniques

Gradient	Resolution	Peak Capacity	Comment
pH (IEF)	ΔpH~0.013	C_{peak}~ 80	Narrow range ampholyte
pH (IEF)	ΔpH~0.034	C_{peak}~ 210	Broad range ampholyte
U (CACE)	$\Delta\mu \sim 0.130 \times 10^{-4}$ cm²/Vs	C_{peak}~ 800	$\langle\mu\rangle \sim 1.0 \times 10^{-4}$ cm²/Vs
E (EFGF)	$\Delta\mu \sim 0.18 \times 10^{-4}$ cm²/Vs	C_{peak}~ 112	$\langle\mu\rangle \sim 1.0 \times 10^{-4}$ cm²/Vs
σ (CGF)	$\Delta\mu \sim 0.034 \times 10^{-4}$ cm²/Vs	C_{peak}~ 210	$\langle\mu\rangle \sim 1.0 \times 10^{-4}$ cm²/Vs
T (TGF)	$\Delta\mu \sim 0.006 \times 10^{-4}$ cm²/Vs	C_{peak}~ 25	$\langle\mu\rangle \sim 1.0 \times 10^{-4}$ cm²/Vs; 40°C

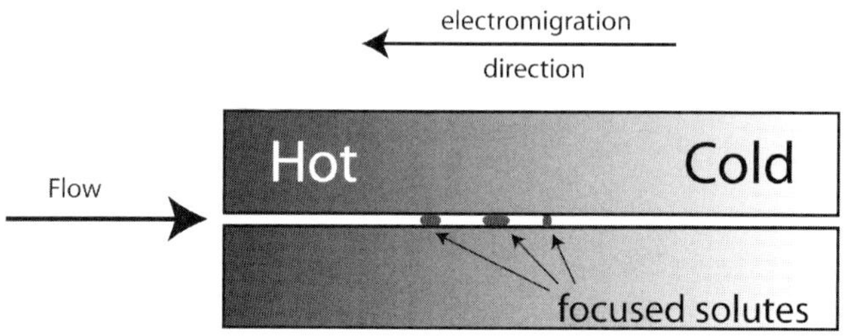

FIGURE 8 TGF uses an applied axial temperature gradient to vary the electric field and focus solutes. This technique, which does not require the use of a membrane or packing, may have some advantages for micro-scale application.

electrophoretic mobilities of the solute (Figure 8). At first glance this approach would appear to be limited by the range of temperatures amenable to the solvent and the solutes, e.g., for proteins ΔT<40°C, but it has the important advantage in that it does not require a membrane in order to focus solutes and so may find application at MEMS scales when deployment of a membrane is difficult.

It is important to note, from Equation (23), that if only temperature-induced variations in the viscosity were considered to provide the focusing gradient, then this approach would not work well since the ratio μ/σ is nearly constant for water between 4°C<T<40°C. However, changing the temperature

$$N_{x,i} = -D_i \frac{dC_i}{dx} + \left(\frac{Q_{in}}{wh} + \left[\frac{\mu_i}{\sigma} \right] \frac{I_{T,x}}{wh} \right) C_i = 0 \tag{23}$$

can also change the buffer conductivity by changing the degree of ionization of the buffer electrolytes (as well as of the target solute) and this can be used as the basis for focusing.

Ross and Locascio[28] have taken advantage of this approach to focus a number of analytes in a microchip format. Introducing a function $f(T)$

in the electrophoretic term which accounts for this shift in the electrophoretic velocity,

$$N_{x,i} = -D_i \frac{dC_i}{dx} + \left(\frac{Q_{in}}{wh} + \left[\frac{\mu_i}{\sigma} \right]_0 \frac{I_{T,x}}{wh} f(T(x)) \right) C_i = 0 \tag{24}$$

where assuming that f is linear in T and that T is linear in x, the velocity function can be expressed as

$$f(T(x)) = x \left[\frac{df}{dT} \right] \left[\frac{dT}{dx} \right] + T_0 \left[\frac{df}{dT} \right] + f_0 \tag{25}$$

with focal point, spatial variance and resolution

$$x_f = -\left(\frac{\frac{Q_{in}}{I_{T,x}} \left[\frac{\sigma}{\mu} \right]_0 + T_0 \left[\frac{df}{dT} \right] + f_0}{\left[\frac{df}{dT} \right] \left[\frac{dT}{dx} \right]} \right) \tag{26a}$$

$$\chi^2 = D_i \left[\frac{\mu_0 I_{T,x}}{\sigma_0 wh} \right] \left[\frac{df}{dT} \frac{dT}{dx} \right]^{-1} \tag{26b}$$

$$R_{VT} = \frac{|\Delta \mu_0|}{4\langle \mu_0 \rangle} \sqrt{\frac{\sigma_0 [Q_{in}/(wh)]^2}{D_i \langle \mu_0 \rangle \frac{I_{T,x}}{wh} \left[\frac{df}{dT} \frac{dT}{dx} \right]}} \tag{26c}$$

Note that since Q_{in} is limited by the electrophoretic velocity of the target solutes, it is coupled to the total current I_T in such a way that resolution increases with $\sqrt{Q}$ rather than Q.

III. RESULTS

Table 1 shows a summary of the theoretical AFM resolutions, which are calculated from the respective formulas for resolution given in the text, and peak capacities, which are estimated by dividing the length of the separation chamber L by 4χ, which is the approximate baseline width of a Gaussian peak. These results have been calculated for a specific set of parameters that pertain to microgram-scale focusing columns and which may not be relevant at different scales or for very different values of the parameters. Yet, these calculations provide a basis for comparison of these techniques and serve as a point of departure for the discussion which follows. The reader should note that, in addition to taking

and the hydrodynamic flow can be varied over the course of the run to resolve and/or elute target solutes.[25,26,31] In its more recent variation called DFGF, where the field is generated by a computer-controlled electrode array,[27] the electric field gradient profile may be manipulated on a point-by-point basis allowing local adjustment of the resolution and peak capacity in different parts of the separation channel. In this configuration, the local resolution can be raised in some parts of the channel to tease apart recalcitrant bands but reduced in other parts where higher resolution is not immediately needed and this can be done at any time during a run. The flexibility of DFGF is not reflected in the calculations of Table 1 since they were based on a static, linear electric field gradient with an electric potential ranging from 1000 V to ground.

E. Grad(σ) Focusing

CGF,[32] which has also been called electromobility focusing,[33,34] exhibits a resolution and peak capacity which are comparable with broad-range IEF under the assumptions used in these calculations. In addition to the restrictions on the voltage, channel geometry, and electrophoretic mobilities, this calculation assumed that the buffer in the purge channel was 1/10 of the conductivity of the buffer introduced into the separation channel and it also assumed that the purge channel was flushed rapidly to maintain that low conductivity everywhere within the purge channel. This produces an exponential variation in the field gradient which makes this technique difficult to compare with the others but, in general, the calculations lead one to expect it would provide performance which is commensurate with the other AFMs.

A couple of points worth noting are that the exponential field profile will provide high peak capacities near the inlet where the gradient is steep and high resolution near the outlet where the gradient is shallow.[33] If desired, the gradient can be made more nearly linear by using a counterflow of low-conductivity buffer in the purge channel at roughly the same flow velocity as in the separation channel, but in the opposite direction. However, if this is implemented, then the purge cannot be used to cool the separation channel since the flow would be too slow to carry away the Joule heat generated in the separation channel, so another means of cooling would have to be implemented.

F. Grad(T) Focusing

Temperature-gradient focusing produced the highest resolving power at the lowest peak capacity. This is the end result of spreading a 40°C temperature excursion over the 10 cm separation channel used in these calculations and, while these numbers could be adjusted by using a different channel length, this result frames the weakness of this technique, i.e., that the permitted temperature rise is limited by the lability of the target solutes. The reader is reminded that all electrophoresis techniques require careful attention to temperature excursions, $\Delta T = T_2 - T_1$, implying that this constraint is really only a matter of degree and will probably only limit its application at larger scales. At capillary, MEMS and smaller scales where large temperature gradients, ∇T, can be attained using small temperature excursions, ΔT, this approach to focusing may have important advantages over some other techniques. The low peak capacity goes hand-in-hand with the exquisite resolving power and should fall in line with the other techniques at small scales where steeper temperature gradients can be applied along the separation channel without producing large temperature excursions.

IV. DISCUSSION

The calculations given above are intended to provide a common basis for comparison of some of the newer AFMs. In addition to this, they illustrate how these types of calculations are performed, how these various techniques could be or have been reduced to practice and, with any luck, have stimulated the reader to consider new AFMs or variations on the ones described here.

The reader is reminded once again that these calculations are based on linearized electrophoresis models whose results may only be meaningful at low solute concentrations where ionic coupling may be ignored even though, in practice, these techniques generally yield solute peaks focused to concentrations in excess of 10 mg/mL where linear models no longer apply. There is a glaring need for *robust* electrophoresis simulations programs and, as numerical packages improve their ability to simulate non-linear electrophoresis, it is hoped that they can provide the tools needed to help explore new concepts in electrokinetics and to accurately design multidimensional, electrokinetic separations systems.

A. Scale

Ampholyte-based IEF finds itself in an odd position relative to scale. While it is used effectively in PAGE[23,35] at mass loadings of 0.1–10 μg, in capillaries and microchips at 0.1–100 ng, and in free solution at 1–100 mg,[36–38] it is unlikely to be used at either larger or smaller scales. The limitation at large-scale stems from the expense of the ampholytes, their tendency to adsorb to proteins and their inability to focus non-amphoteric compounds. At the extreme small scales, e.g., subpicogram, needed for the final separation dimension(s) of an integrated proteomics nanochip, placement or delivery of the ampholytes presents an unnecessary complication since other media, e.g., *Immobilines®*, or the alternative methods described here could be used in a way that avoids these problems. Ampholyte-based IEF is also limited by the practical pH range of available ampholyte fractions. In free solution, operation outside the 3–10 pH range is challenging but high-resolution IEF with $\Delta pH < 0.5$ is virtually impossible with commercially available ampholytes.

This is not the case with the AFMs because the gradients used here are not fundamentally limited by chemistry, as is the case with commercial ampholytes, or by "delivery" constraints since they can be designed into separation microchannels at the location where they will be used. The alternative systems can, in principle, be designed with extremely shallow gradients having resolving power far in excess of ampholyte-based IEF. Furthermore, the AFM channels can be fabricated to include a pattern of gradients that are optimized for a given mixture or engineered for general-purpose use, e.g., a steep gradient for collecting peaks followed by a shallow gradient for resolving peaks. The only exception to this is temperature-gradient focusing which, for proteins, is limited to about $\Delta T \sim 40°C$ for the steep gradients used to collect peaks but can still produce extremely shallow gradients.

Many of the AFMs are amenable to increased scale. Work along these lines has already started for DFGF[27] and will likely be extended to conductivity- and velocity-gradient focusing over the next few years. For obvious reasons alluded to above, temperature-gradient focusing seems an unlikely candidate for large-scale processing but has been demonstrated at MEMS scales and may have important advantages at that scale. Finally, it is my opinion that CACE, and any other electrochromatography technique that relies on a granular packing for separation, will be difficult to bring to process scales, even with internal cooling coils.[7] At small scales, the electrochromatographies are challenged by packing issues, but these can be circumvented by using derivatized nanochannels, monoliths or *in situ* fabrication.

B. Integrated Networks

The future of complex biochemical analyses, e.g., systeomics, lies in the integrated MEMS networks commonly referred to as "lab-on-a-chip." These chips will have to accept a "real-world" sample in, at the very least, 1 μL volume, pull out one or more targets and prepare them for detection. Low-abundance targets will have to be concentrated to produce an adequate signal at the detector and this requires that they be moved from a 1 μL volume down, perhaps, to a sub-picoliter volume, according to their relative abundance in the starting sample.

Integration will allow a chip to apply a variety of unit operations, e.g., fractionation, reaction, dialysis, digestion, etc., to the sample components as they pass through the process cascade on their way to the detector(s) but conventional wisdom currently holds that the HPLCs will not scale well below $L = 1$ mm because column length is required to resolve peaks. This stricture does not apply to solid-phase extraction or other on-off or "step" affinity chromatographies, but these techniques may be hampered by placement/delivery concerns if a packed bed is used.

Although it is not immediately obvious from Equation (10), the resolving power of IEF is largely independent of channel length if the end-to-end voltage drop across the channel is maintained as channel length is reduced. Since the electric field increases in inverse proportion to the channel length and the power density goes as the square of the field, the key design issue that arises is the temperature excursion but, if the effective channel diameter is reduced with E or faster, temperature excursions should not be a problem. These same properties appear to hold for the AFMs although again, this is not obvious from the formulas given for the various resolutions. However, if one notes that the axial flow in the channel and the current must be proportional to the electric field and that the electric field is inversely proportional to the channel length if the voltage drop is held constant, then the channel length drops out of each of these formulas just as it does for IEF. This suggests that any of these electrokinetic focusing techniques could be used throughout the separation cascade as the channel diameters are reduced to accept the smaller volumes of concentrated, low-abundance solutes.

V. CONCLUSION

While the development of AFMs radiates from O'Farrell's invention of CACE,[4] it is firmly grounded in the EGMs described in Giddings' classic paper.[1] To date only a fraction of the possible variants of the alternative focusing techniques have been tested in the laboratory and all of these involve only a single field gradient along a single field-flow axis and using only electric fields and flows as the counteracting forces. In the near future, it is expected that this work will expand to include binary and then tertiary interacting field gradients followed shortly thereafter by multidimensional (spatial) gradients applied over electrode (or thermal, etc.) arrays, multidimensional (temporal) gradients and, eventually, multiple separation dimensions using orthogonal focusing methods.

ACKNOWLEDGMENTS

This material is based upon work supported in part by the National Science Foundation under Grants DBI-0096745 and BES-9970972.

REFERENCES

1. Giddings, J. C. and Dahlgren, K. Resolution and peak capacity in equilibrium—gradient methods of separation. *Sep. Sci.* 6(3):345–356, 1971.
2. Giddings, J. C. *Unified Separation Science*. John Wiley and Sons, Inc., New York, 1991.
3. O'Farrell, P. H. Method of Isotope Enrichment, US patent 4290855 (1981).
4. O'Farrell, P. H. Separation techniques based on the opposition of two counteracting forces to produce a dynamic equilibrium. *Science* 227(4694):1586, 1985.
5. Hunter, J. B. *Progress in Mathematical Modelling of CACE*, Vol. 23, Marcel Dekker, NY, p. 875, 1988.
6. Locke, B. R. and Carbonell, R. G. A theoretical and experimental study of counteracting chromatographic electrophoresis. *Sep. and Purif. Methods* 18(1):1–64, 1989.
7. Gobie, W. A. and Ivory, C. F. Continuous counter-acting chromatographic electrophoresis, *Biotechnol. Prog.* 6(1):21–32, 1990.
8. Ivory, C. F. A brief review of alternative electrofocusing techniques, *Sep. Sci. Technol.* 35(11):1777–1793, 2000.
9. Mosher, R. A., Saville, D. A. and Thormann, W. *The Dynamics of Electrophoresis*. VCH Publishers, Inc., New York, 1992.
10. Mosher, R. A., Gebauer, P., Caslavska, J. and Thormann, W. Computer-simulation and experimental validation of the electrophoretic behavior of proteins. 2. Model improvement and application to isotachophoresis. *Anal. Chem.* 64(23):2991–2997, 1992.
11. Mosher, R. A., Gebauer, P. and Thormann, W. Computer-simulation and experimental validation of the electrophoretic behavior of proteins. 3. Use of titration data predicted by the proteins amino-acid-composition. *J. Chromatogr.* 638 (2):155–164, 1993.
12. Mosher, R. A. and Thormann, W. High-resolution computer simulation of the dynamics of isoelectric focusing using carrier ampholytes: the post-separation stabilizing phase revisited. *Electrophoresis* 23(12):1803–1814, 2002.
13. Thormann, W., Huang, T., Pawliszyn, J. and Mosher, R. High-resolution computer simulation of the dynamics of isoelectric focusing of proteins. *Electrophoresis* 25(2):324–337, 2004.
14. Cao, C. X., Zhou, S. L., Qian, Y. T., He, Y. Z., Wang, C. R., Zheng, X. Y. and Chen, W. K. Investigations on factors that influence the moving neutralization reaction boundary method for capillary electrophoresis and isoelectric focusing. *J. Chromatogr. A* 952(1–2):29–38, 2002.
15. Arnaud, I. L., Josserand, J., Rossier, J. S. and Girault, H. H. Finite element simulation of off-gel (TM) buffering. *Electrophoresis* 23(19):3253–3261, 2002.
16. Kim, S. H., Otani, R., Shirai, W., Shibasaki-Kitakawa, N., Kitakawa, A. and Yonemoto, T. Mathematical modeling of continuous separation process of proteins using partial recycling free-flow isoelectric focusing with feed supply and product withdrawal. *Kagaku Kogaku Ronbunshu* 27(2):174–180, 2001.
17. Cabrera, C. R., Finlayson, B. and Yager, P. Formation of natural pH gradients in a microfluidic device under flow conditions: model and experimental validation. *Anal. Chem.* 73(3):658–666, 2001.
18. Taylor, G. I. Dispersion of soluble matter in solvent flowing through a tube. *Proc. Roy. Soc. A* 219:186–203, 1953.
19. Aris, R. On the dispersion of a solute in a fluid flowing through a tube. *Proc. Roy. Soc. A* 235:67–77, 1956.
20. McCoy, B. J. Counteracting chromatographic electrophoresis and related imposed–gradient separation processes. *AIChE J.* 32(9):1570, 1986.
21. Boyce, W. E. and DiPrima, R. C. *Elementary Differential Equations and Boundary Value Problems*, 3rd ed., John Wiley and Sons, Inc., New York, 1977.
22. Abramowitz, M. and Stegun, I. *Handbook of Mathematical Functions*, Dover Publications, Inc., New York, 1965.

23. Righetti, P. G. *Isoelectric Focusing: theory, methodology and applications*, Elsevier Biomedical Press, Amsterdam, 1983.
24. Righetti, P. G., Faupel, M. and Wenisch, E. Preparative electrophoresis with and without immobilized pH gradients, In *Advances in Electrophoresis* (Chrambach, A., Dunn, M. J. and Radola, B. J. Eds.) VCH, New York, pp. 159–200, 1992.
25. Koegler, W. S. and Ivory, C. F. Focusing proteins in an electric field gradient. *J. Chromatogr.* 229:229–236, 1996.
26. Koegler, W. S. and Ivory, C. F. Field gradient focusing: a novel method for protein separation. *Biotechnol. Prog.* 12(6):822–836, 1996.
27. Huang, Z. and Ivory, C. F. Digitally controlled electrophoretic focusing. *Anal. Chem.* 71(8):1628–1632, 1999.
28. Ross, D. and Locascio, L. E. Microfluidic temperature gradient focusing. *Anal. Chem.* 74(11):2556–2564, 2002.
29. Balss, K. M., Vreeland, W. N., Howell, P. B., Henry, A. C. and Ross, D. Micellar affinity gradient focusing: a new method for electrokinetic focusing. *J. Am. Chem. Soc.*, ASAP Web Release (Communication), 2004.
30. Giddings, J. C., Yang, F. J. F. and Myers, M. N. Flow field–flow fractionation: a versatile new separation method, *Science* 193:1244–1245, 1976.
31. Wang, Q. G., Lin, S. L., Warnick, K. F., Tolley, H. D. and Lee, M. L. Voltage-controlled separation of proteins by electromobility focusing in a dialysis hollow fiber. *J. Chromatogr. A* 985(1–2):455–462, 2003.
32. Greenlee, R. D. and Ivory, C. F. Protein focusing in a conductivity gradient. *Biotechnol. Prog.* 14(2):300–309, 1998.
33. Tolley, H. D., Wang, Q. G., LeFebre, D. A. and Lee, M. L. Equilibrium gradient methods with nonlinear field intensity gradient: a theoretical approach. *Anal. Chem.* 74(17): 4456–4463, 2002.
34. Wang, Q. G., Tolley, H. D., LeFebre, D. A. and Lee, M. L. Analytical equilibrium gradient methods. *Anal. Bioanal. Chem.* 373(3):125–135, 2002.
35. Deutscher, M. P. Guide to Protein Purification. In *Methods in Enzymology* (Abelson, J. N. and Simon, M. I. Eds.) Academic Press, New York, 1990.
36. Bier, M. Rotating apparatus for isoelectric focusing. US patent 4588492 4588492, 1986.
37. Lubman, D. M., Kachman, M. T., Wang, H. X., Gong, S. Y., Yan, F., Hamler, R. L., O'Neil, K. A., Zhu, K., Buchanan, N. S. and Barder, T. J. Two-dimensional liquid separations-mass mapping of proteins from human cancer cell lysates. *J. Chromatogr. B* 782(1–2):183–196, 2002.
38. Wall, D. B., Kachman, M. T., Gong, S. Y. S., Parus, S. J., Long, M. W. and Lubman, D. M. Isoelectric focusing nonporous silica reversed-phase high-performance liquid chromatography/electrospray ionization time-of-flight mass spectrometry: a three-dimensional liquid-phase protein separation method as applied to the human erythroleukemia cell-line. *Rapid Commn. Mass Spectrom.* 15(18):1649–1661, 2001.

INDEX